Rare Earth Nanotechnology

I humbly dedicate this book to my parents,
Mr Tan Boon Chew and Mdm Lee Ah Sioh,
and sister, Ms Tan Foong Yee,
who have provided the foundation and
compassion on which I have been blessed
to rely and build.

Contents

Preface

Rare earth elements (REE) are gaining ubiquitous importance in modern technology and have been touted as the "vitamin of chemistry." They help technologies perform better and have their own unique characteristics. Many high-technology industries depend heavily on these unique elements for the manufacture of permanent magnets and batteries, which are vital to efficient military and green technologies such as wind turbines and hybrid batteries, as well as in smartphones and laptops.

REE are, in fact, not rare, and most of them are fairly abundant in the earth's crust. What is rare about REE is their supply. China controls 95% of the world's REE production, not exactly due to geological luck but more due to economic and scientific strategies. In 2011, REE attracted unprecedented news when China announced a 70% cut in its rare earth production, sending shockwaves through the world as it feared a supply crunch. Since then, various plans to establish rare earth production outside China have been in the pipeline, with Australian mining company Lynas having been successfully granted approval to build one of the largest rare earth refineries in Malaysia.

This book was conceived prior to these events, when the exploration, research, and development of rare earth materials in nanotechnology were burgeoning at the start of the millennia, especially in the fields of nanomedicine and nanophosphors. This book, therefore, focuses on the potential applications of rare earth materials in these areas and their state of the art in these applications. The aim of each chapter is to review and highlight the strategies and insights of the research work in the relevant areas, in a hope to establish continued and long-term research efforts of these amazing materials in nanotechnology.

This book consists of six chapters put together in a cohesive and sequential manner, but they can be read as standalone chapters. As the properties of rare earth elements can be found in many textbooks and journal articles, they have not been included in this book. The book begins with highlighting key strategies in the design and synthesis of various types and forms of rare earth nanomaterials (Chapter 1), followed by Chapter 2, which discusses various approaches to synthesizing rare earth nanomaterials of different morphologies and their surface modification to render them suitable for their intended applications. Rare earth materials have intriguing optical and magnetic properties. In Chapter 3, recent works on the application of rare earth nanoparticles in fluorescence microscopy are highlighted, with a strong focus on upconversion rare earth nanoparticles as they are most suited as imaging probes for biological specimen. In the pursuit of better imaging contrast to achieve more accurate diagnosis, there has been much interest, and success, in the use of rare earth nanoparticles as "proof-of-principle" magnetic resonance imaging contrast agents. Their state of the art is discussed in Chapter 4. Chapter 5 demonstrates the foresight of researchers for bimodal contrast agents in bioimaging technology, of which the optical and magnetic properties of rare earth nanomaterials are simultaneously exploited to achieve more accurate and sensitive imaging in fluorescence and magnetic resonance imaging. The last chapter presents the advances and promises of rare earth nanomaterials as cheaper and more efficient lighting materials in light-emitting diodes, resonating the global need for green lighting technologies.

Rare earth elements will continue to exert their significant impact in modern technologies in the coming decade. Supply–demand will shift toward equilibrium with the discovery of more mines and the construction of new refineries.

Hopefully, this book will provide the readers, be it researchers, engineers, or policymakers, with bountiful ideas and inspirations to effect a new level of nanotechnological revolution using REE, especially in the much-needed energy and healthcare sectors.

Timothy T.Y. Tan

Chapter 1

Synthesis of Rare Earth Nanomaterials

Chun-Hua Yan, Chao Zhang, and Ling-Dong Sun
Beijing National Laboratory for Molecular Sciences, State Key Laboratory of Rare Earth Materials Chemistry and Applications, PKU-HKU Joint Laboratory in Rare Earth Materials and Bioinorganic Chemistry, Peking University, Beijing, 100871, China
yan@pku.edu.cn

1.1 Introduction

This chapter discusses the synthesis of rare earth nanomaterials, mainly covering rare earth oxides, sulfides, halides, and oxysalts. Basically, the synthetic routes of rare earth nanomaterials can be sorted into two main categories: the "dry" methods and the "wet" methods. The former refers to the synthetic routes like solid-state reaction, self-propagating synthesis, and several physical routes (e.g., CVD, PLD, and magnetic sputtering). In contrast, the wet methods generally involve solution-based processes, during which various parameters can be finely adjusted (including reaction time, temperature, concentration, pH value, as well as the utilization of coordination reagents, templates, mineralizers, etc.), and thus exhibit particular superiorities in controlling phase purity, chemical homogeneity, size, and morphology of the final products. In addition,

Rare Earth Nanotechnology
Edited by Timothy T.Y. Tan

ISBN 978-981-4316-30-9 (Hardcover), 978-981-4364-20-1 (eBook)
www.panstanford.com

compared to "dry" methods, which commonly resort to rigorous conditions such as high temperature, high pressure, or high vacuum, the "wet" syntheses are usually carried out under relatively mild conditions, which also help to lower energy consumption and total cost. Therefore, the discussion in the following sections will be mainly focused on wet methods, including precipitation, sol-gel, hydrothermal/solvothermal, thermal decomposition, microwave, microemulsion routes, and so on.

1.2 Precipitation/Co-Precipitation Route

The precipitation/co-precipitation route usually affords products with pure phase, and the experimental procedures are relatively simple. The metal ions are first precipitated from solutions as hydroxides, carbonates, and oxalates, which undergo the subsequent calcination treatment to form the products. Binary compounds, such as CeO_2 and REF_3, can be readily obtained by this method; yet for those complex systems (doped systems and ternary systems, for example), the precipitation procedure requires particular care because the precipitation rates can vary for different metal ions. In such cases, certain coordination reagents may be necessary to adjust the synchronicity in the subtle co-precipitation procedure, so as to obtain homogeneous products with predesigned compositions. Due to the relatively simple operations involved in this route, mass production is easily achieved.

The rare earth compound ceria (CeO_2) is currently under most extensive and intensive investigation. It adopts a cubic fluorite phase in a wide temperature range (from ambient temperature up to its melting point), and the fluorite structure can be preserved to a considerable extent under reductive atmospheres. The reduction from Ce^{4+} ions to Ce^{3+} ions can generate oxygen vacancies, which act as highly reactive sites for plenty of catalysis redox reactions. When the size of ceria is reduced down to the nanometer dimensions, the catalytic activity is much elevated due to the enlarged surface area, enhanced oxygen storage capacity, which caters to the demands of three-way catalysts, fuel cells, and so on. Due to the cubic phase, nanoceria tends to expose low-index crystal surfaces, i.e., {100}, {110}, and {111}, and usually takes the shape of nanocubes, nanooctahedra, nanowires, and nanotubes.

Ceria nanocrystals can be readily prepared by the precipitation method, using either Ce(IV) or Ce(III) soluble salts as Ce source. Nanoceria can be directly obtained in aqueous solution, and in some cases, intermediates like hydroxides or carbonates are obtained first, which form nanoceria through post-heating treatment in air.

Zhang *et al.* reported the synthesis of ceria using $Ce(NO_3)_3$ as Ce source and hexamethylenetetramine as the base [1]. Particles in the size range of 3–12 nm were prepared at room temperature, and larger particles were made by calcination at 400–800°C. Han *et al.* obtained ceria nanotubes via two successive stages: precipitation and aging (Fig. 1.1) [2]. Ammonia was used to form precipitates from aqueous solution of $Ce(NO_3)_3$ at 100°C, and the precipitates were aged for a long time of 45 days. The formation of the tubular structure was found to be strongly dependent on the precipitation temperature and aging time.

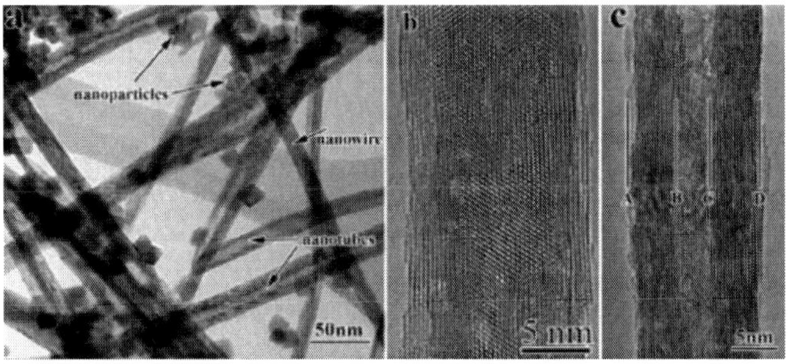

Figure 1.1 (a) Typical morphology of the ceria samples. There are three kinds of nanostructures: nanoparticles, nanowires, and nanotubes as marked in the figure. (b) High-resolution image of a nanowire. (c) High-resolution image of a nanotube.

Du *et al.* prepared ceria nanorods along the [211] or [110] direction by refluxing at 100°C, and pearl-chain-like nanostructures at lower temperatures (70–90°C) (Fig. 1.2) [3]. An oriented attachment mechanism was proposed, claiming that the nanorods and pearl chains are formed by self-organization of truncated octahedral ceria nanocrystals.

Ceria–zirconia solid solution, as a well-known ceria-based material with large oxygen storage capacity, is also extensively studied. Hydroxide co-precipitation synthesis of $Ce_{1-x}Zr_xO_2$ was reported by Deshpande *et al.* [4]. Cerium ammonium nitrate and zirconyl chloride were used as Ce and Zr source, respectively, and nitric acid as the peptizing agent. They also employed the nanocasting technique to prepare a range of mesoporous $Ce_{1-x}Zr_xO_2$ beads, using polymeric porous beads as the hard template [5].

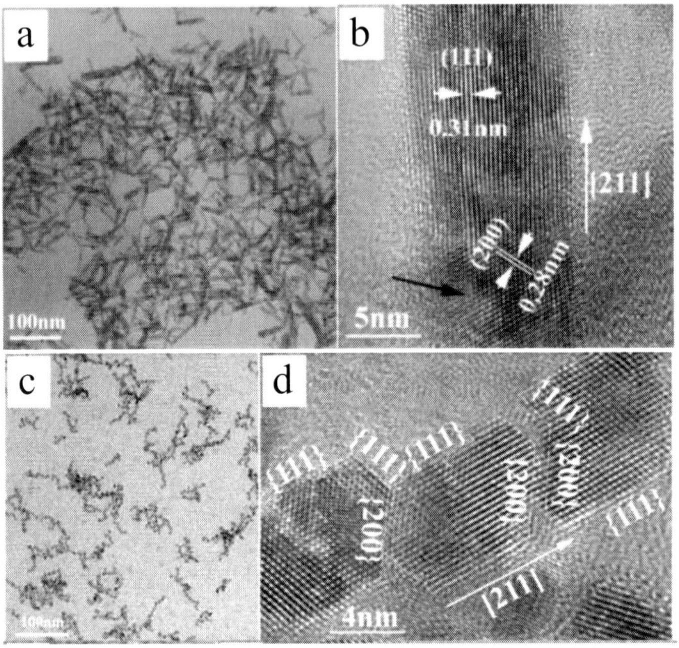

Figure 1.2 (a, b) TEM and HRTEM images of the ceria nanorods; (c, d) TEM and HRTEM images of the ceria pearl-chain-like nanostructures.

Rare earth sesquioxides (RE_2O_3) are a class of materials of great importance in catalysis and luminescence fields. Wakefield *et al.* developed a colloidal precipitation route to prepare Eu_2O_3 and Y_2O_3:Eu nanoparticles [6, 7]. The luminescent properties of these materials were also examined. Oxide nanotubes of Er, Tm, Yb, Lu were synthesized by Yada *et al.* by the method of homogeneous precipitation with dodecylsulfate assemblies as templates [8]. These nanotubes can be further tailored to compose hierarchical 2D and 3D microstructures [9].

Rare earth fluorides generally exhibit good thermal stabilities and high ionic nature, and find applications primarily in solid-state lasers, lighting, and display. Pure and doped rare earth fluorides can be readily obtained by the co-precipitation method. van Veggel *et al.* reported the preparation of LaF_3:R^{3+} (R = Eu, Er, Nd, and Ho) nanoparticles from rare earth nitrates and NaF in ethanol/water mixed solvent. A capping ligand, namely, ammonium di-*n*-octadecyldithioposhate, was used to prevent the particles from agglomeration, and the as-obtained nanoparticles can be easily redispersed in organic solvents [10]. Chow *et al.* also employed the co-precipitation method to prepare multicolor upconversion luminescent LaF_3 nanocrystals doped with Yb^{3+} as sensitizer ions and Er^{3+}, Ho^{3+}, Tm^{3+} as the activator ions [11]. Chen *et al.* synthesized EuF_3 nanocrystals with different morphologies, including nanoplates, nanospheres, nanobundles, nanorods, and nanowires (Fig. 1.3) [12, 13]. Many different fluoride sources (HF, NaF, KF, NH_4F, RbF, CsF, and $NaBF_4$) were used to precipitate the rare earth ions. Similar routes were employed by many research groups, affording different REF_3 nanoparticles with varied morphologies [14–18].

Figure 1.3 SEM images of (a) hexagonal, (b) orthorhombic EuF_3 nanocrystals, (c) EuF_3 nanospindles and (d) EuF_3 nanodisks.

In addition to fluorides, alkali-rare-earth complex fluorides (denoted as $AREF_4$) are also an important class of rare earth materials, which have attracted extensive attention for their applications in laser, display, and bioimaging. $AREF_4$ nanoparticles can be readily obtained by the co-precipitation method. Chen *et al.*

prepared hexagonal $NaEuF_4$ and cubic $Na_5Eu_9F_{32}$ nanocrystals by simply reacting $Eu(NO_3)_3$ and NaF in water [19]. They found that the bundle-like EuF_3 nanostructures were yielded first, which then reacted with NaF to give $NaEuF_4$ with one-dimensional morphology. Chen *et al.* prepared nearly monodisperse $NaYF_4$:Yb,Er nanoparticles by co-precipitation of Y^{3+}, Yb^{3+}, Er^{3+} with NaF. EDTA was introduced as the chelating reagent, and by varying the EDTA/RE^{3+} ratio, particle size could be effectively controlled [20]. Karbowiak *et al.* synthesized cubic $KGdF_4$ and $KGdF_4$:Eu^{3+} nanocrystals using soluble rare earth chlorides and NH_4HF_2 as starting materials based on co-precipitation process [21, 22].

Among various rare earth oxysalts, rare earth orthophosphates and orthovanadates are of particular importance. These compounds generally exhibit low water solubility, high thermal stability, and high luminescent efficiency. In particular, the vanadate matrix shows a strong charge-transfer absorption band located at 200–350 nm, which facilitates the energy migration from the host matrix to rare earth ions. This can significantly enhance the absorption and excitation efficiency of the doped luminescent rare earth ions. The above advantageous characteristics of these materials guarantee their applications as phosphors and sensors in display, lighting, and bioprobing. A typical precipitation route for rare earth phosphates and vanadates usually begins with soluble salts containing corresponding cations and anions, for example, rare earth nitrates and chlorides, and phosphate, phosphoric acid, and vanadates. The pH value of the system always plays a crucial role during the whole synthetic process, especially for vanadates because vanadium(V) exists in different forms, such as VO_4^{3-} monomers or oligomers like $V_3O_9^{3-}$ and $V_{10}O_{28}^{6-}$ anions, at varied acidities. Gao *et al.* reported pure and doped hexagonal $LaPO_4$ nanorods with typical dimensions of 8 nm in diameter and 80 nm in length [23]. The nanorods were prepared from NaH_2PO_4 and $LaCl_3$ aqueous solutions at 100°C heated by an oil bath. For heavy rare earth (Ho–Lu) and Y, the reaction temperature can be even lower. Di *et al.* synthesized YPO_4:Eu nanowires at a temperature as low as 70°C [24]. Buissette *et al.* reported the colloidal synthesis of sub-10-nm $LaPO_4$:Ce,Tb and $LaPO_4$:Eu nanoparticles by aging a mixed solution of rare earth nitrates/chlorides and sodium tripolyphosphate ($Na_5P_3O_{10}$, TPP) at 90°C for 3 h [25]. TPP acted as both the source of orthophosphate anions and the complexing reagent.

Huignard *et al.* exploited the room temperature co-precipitation method using $Y(NO_3)_3$, $Eu(NO_3)_3$, and $Na_3VO_4/NaVO_3$ as starting materials and obtained *t*-YVO_4:Eu^{3+} nanoparticles with sizes around 15–30 nm [26]. The as-prepared YVO_4:Eu^{3+} nanoparticles could be further stabilized into a colloidal solution by introducing sodium hexametaphosphate. The Eu^{3+} quenching concentration was found to be elevated, and luminescence efficiency reduced for nanosized YVO_4:Eu^{3+}. This was probably due to the nonradiative de-excitation pathways resulting from the surface defects. Huignard *et al.* reported that when competitive chelating ligands like citrate anions were introduced, the growth of $REVO_4$ nanoparticles can be tuned in a more delicate manner, probably due to the competition between coordination and precipitation of rare earth ions [27]. Isobe *et al.* doped Bi^{3+} along with Eu^{3+} ions into the *t*-YVO_4 lattice so as to exploit the excitation energy of irradiation light 300–400 nm in wavelength [28]. The co-doped Bi^{3+} could serve as sensitizer for Eu^{3+} through the energy transfer from Bi^{3+} 6s orbital to V^{5+} 3d orbitals. Bismuth(III) citrates were used rather than nitrates for the strong hydrolysis tendency of the latter. van Veggel *et al.* prepared *t*-$LaVO_4$ nanoparticles with various rare earth dopants (Eu, Tm, Nd, Er, Ho, Dy, Sm, Pr) in ethanol/water media at 75°C [29]. The co-precipitation reaction was performed in the presence of surfactant $NH_4(n\text{-}C_{18}H_{37}O)_2PS_2$. The dithiophosphate anion ligands were found to help to control the growth of $LaVO_4$ nanoparticles. As a representative, *t*-$LaVO_4$:Eu^{3+} nanoparticles were obtained in irregular shapes with a size around 6–10 nm. These nanoparticles can be well dispersed in nonpolar solvents due to the hydrophobic ligands attached to the surface.

Other rare earth oxysalts can also be prepared via the co-precipitation method, for example, perovskite-structured $LaAlO_3$ [30], pyrochlore-structured $Y_2Ti_2O_7$, $La_2Sn_2O_7$, and $La_2Zr_2O_7$ [31, 32], monoclinic and cubic $La_2Mo_2O_9$, and $La_2W_2O_9$ [33, 34], and perovskite-structured rare earth transition metal (M) complex oxide nanoparticles (M = Mn, Fe, Co, Ni, and Cu) [35–37]. Owing to the limited pages, these works will not be discussed in details.

1.3 Sol-Gel Route

In a typical sol-gel route, precursors are mixed in solution and then hydrolyzed and polymerized to form a (meta)stable sol system, which subsequently undergoes a gelation process. The resultant gel, consisting of either crystalline or amorphous phases, is subsequently heated to yield the required products. As the sol-formation process is relatively slow and carried out under ambient conditions, this method allows the elaborate control over the elemental stoichiometry and the chemical homogeneity of the final products. In particular, when surfactant reagents are introduced into the initial solution to form self-assembly entities (micelles and vesicles, for instance) as soft templates, porous materials can be expected, which find applications in catalysis, fuel cells, photovoltaics, and so on. In addition, the sol-gel route can be easily combined with other techniques, for example, casting and electrospinning, to fabricate films and fibers.

Brezesinski *et al.* demonstrated a concept to fabricate mesoporous crystalline ceria thin films based on evaporation-induced self-assembly (EISA) and subsequent heat treatment, using a suitable block copolymer and an ionic liquid (IL) as templates[38]. Ceria–zirconia solid solution can also be prepared via sol-gel routes. Thammachart *et al.* employed urea as the hydrolysis catalyst to synthesize highly uniform nanoparticles of Ceria–zirconia [39]. Yan *et al.* reported ordered mesoporous $Ce_{1-x}Zr_xO_2$ solid solutions prepared with the assistance of block copolymer as the surfactant and soft template (Fig. 1.4) [40]. The product shows uniform mesopores in hexagonal arrangement, with crystalline walls.

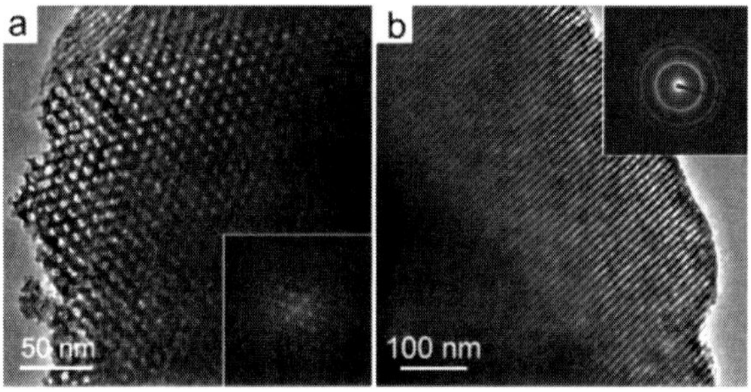

Figure 1.4 TEM images of the mesoporous $Ce_{1-x}Zr_xO_2$ (x = 0.5) recorded along the (a) [001] and (b) [110] orientations. The inset in (a) is the corresponding FFT (fast Fourier transform) diffraction image, and the one in (b) is the corresponding SAED pattern.

The sol-gel route also shows robust capability in synthesizing RE_2O_3 nanomaterials, including Y_2O_3:Tb, Y_2O_3:Eu [41, 42]. Wu *et al.* combined the sol-gel technique with hard template approach and fabricated Eu_2O_3 nanotubes using anodic aluminum oxide [43].

Hou *et al.* prepared Eu^{3+}, Sm^{3+}, and Dy^{3+} doped *t*-$YP_{0.8}V_{0.2}O_4$ nanofibers and *t*-YVO_4 microbelts by exploiting the sol-gel process and electrospinning technique [44]. The sol precursors were mechanically molded into one-dimensional shape and then crystallized by post-annealing treatment.

As for other rare earth oxysalts, the sol-gel method exhibits robust viability and hence related reports are abundant. Herein, only a few representative works will be discussed. Hreniak *et al.* reported $LaAlO_3$:Eu^{3+} nanoparticles prepared via sol-gel derived Pechini method, and an enhancement of luminescence lifetimes with decreasing of the particle sizes was found [45]. Fujihara *et al.* prepared $Y_2Sn_2O_7$: Eu^{3+} thin films with coordination reagent of citric acid and cross-linking reagent of poly(ethylene glycol) [46]. The sol was dip-coated on quartz glass substrate and then calcined at 800–1000°C for 1 h, yielding a film with the thickness of approximately 150 nm, which displays orange-red emissions upon ultraviolet excitation. Li *et al.* reported pyrochlore-structured $La_2Zr_2O_7$ nanofibers prepared by electrospinning technique (Fig. 1.5) [47]. PVP, lanthanum nitrate, and zirconium oxychloride were used as precursors, which were then calcined at 1000°C for 12 h, affording nanofibers with a diameter of 100–500 nm. Subramania *et al.* prepared Pr^{3+}-doped $La_2Mo_2O_9$ by pyrolysis of polyacrylate salt precursors via in situ polymerization of metal salts and acrylic acid [48]. Popa *et al.* prepared pure single-phase $LaMeO_3$ (Me = Mn, Fe, Co) nanopowders with homogeneous microstructure [49–52]. Niu *et al.* synthesized a series of $RFeO_3$ (R = Nd, Sm, Eu, Gd, Dy, Ho,) perovskites based on the sol-gel method and investigated their performances in gas-sensing (including H_2, H_2S, CO, etc.) and photocatalytic degradation of several water-soluble dyes [53–58].

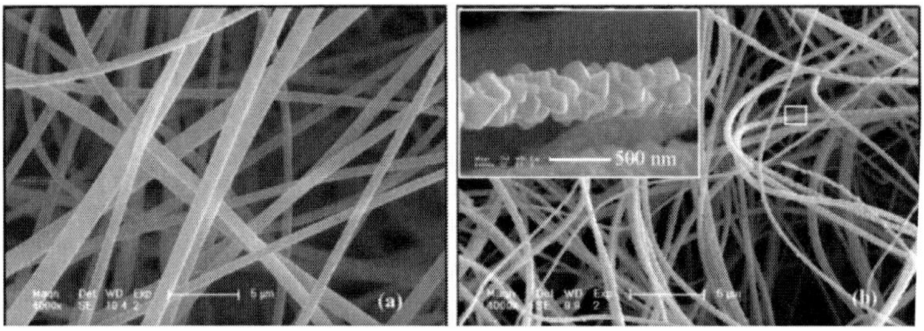

Figure 1.5 SEM micrographs for (a) PVP-precursor fibers and (b) $La_2Zr_2O_7$ fibers calcined at 1000°C.

1.4 Hydrothermal/Solvothermal Route

The hydrothermal/solvothermal synthesis is typically carried out in autoclaves, where the superheated solvents and the autogenerated high pressure provide a microenvironment distinctly different from those in common synthetic systems. As a result, the products usually exhibit few defects, high crystallinity, fine homogeneity, and sometimes special morphologies and thermodynamically metastable phases, especially when appropriate chelating reagents or mineralizers are employed.

Cerium hydroxyl carbonate ($Ce(OH)CO_3$) can be readily produced by hydrothermal reactions and thus frequently used as intermediates for CeO_2. Li *et al.* obtained $Ce(OH)CO_3$ through hydrothermal route using cerium oxalate as the Ce source, and ceria was obtained by subsequent sintering [59]. Similarly, Lu *et al.* prepared spindle-like orthorhombic $Ce(OH)CO_3$ nanocrystals through a hydrothermal synthesis with urea at 100°C [60]; Han *et al.* prepared flower-like nanorod bundles of hexagonal $Ce(OH)CO_3$ at 200°C [61]; Chen *et al.* reported the solvothermal synthesis of $Ce(OH)CO_3$ and CeO_2 in mixed solvents of water and ethanol [62]. Du *et al.* reported the hydrothermal synthesis of uniform triangular plate-like and shuttle-like $Ce(OH)CO_3$ assisted by different surfactants (CTAB or PVA) [63, 64]. The intermediates were converted into ceria by calcination in air. Sun *et al.* reported the synthesis of nearly monodisperse flower-like CeO_2 microspheres consisting of nanosheet as petals [65]. $Ce(OH)CO_3$ spheres with similar morphology were prepared first as an intermediate, using glucose, acrylamide, and ammonia to form a gel, which was then subjected to hydrothermal treatment at 180°C. An ensuing two-step calcination procedure affords the final products (Fig. 1.6).

Chen *et al.* developed an EDTA-assisted hydrothermal route for the selective preparation of submicrorods of CeO_2 via an incomplete reaction [66]. They also found that uniform nanoparticles were formed through a complete reaction between $NaClO_3$ and Ce–EDTA complexes. Yan *et al.* used $Na_3PO_4 \cdot 6H_2O$ as a mineralizer in a hydrothermal process at 170°C to synthesize ceria nanocrystals [67]. During the reaction, octahedral-shaped ceria particles formed first, and then underwent a phosphate-ion-related dissolution–recrystallization process to form nanorods.

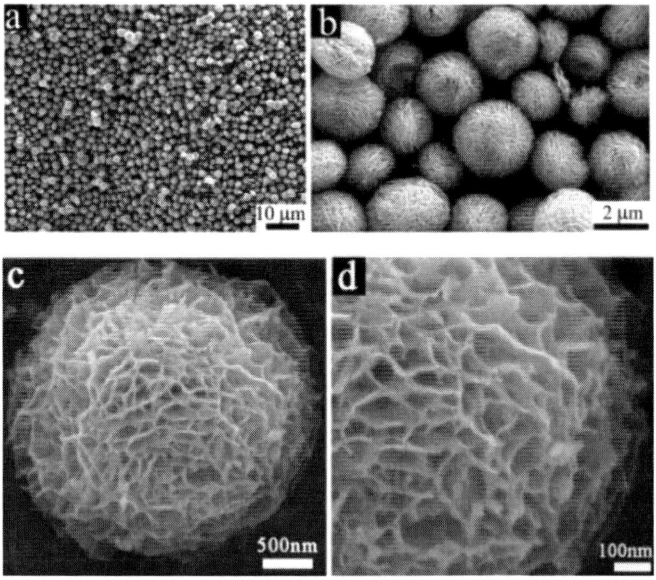

Figure 1.6 SEM images of the ceria microspheres: (a, b) overall morphology; (c, d) high-magnification SEM images of an individual sphere, revealing the microstructures therein.

Yan *et al.* reported the formation of nanopolyhedra, nanorods, and nanocubes of CeO_2 through a facile hydrothermal route (Fig. 1.7) [68]. The experiments were carried out at temperatures in the range of 100–180°C under different NaOH concentrations. Single crystalline ceria nanorods of 10 nm × (50–200) nm with a growth direction of [110] are obtained at 100°C with over 6 mol L^{-1} NaOH, by virtue of the temporary formation of an intermediate of hexagonal $Ce(OH)_3$.

Tang *et al.* reported the synthesis of ceria nanotubes using $Ce(OH)_3$ as the intermediate. Hydrothermal reaction under oxygen-free conditions was used to obtain $Ce(OH)_3$ nanotubes [69]. Subsequently, CeO_2 nanotubes were prepared by annealing the $Ce(OH)_3$ nanotubes at 450°C under a mild controlled condition, i.e., using reducing atmosphere instead of air. The temperature was raised slowly to allow the smooth structure modification during the conversion. Zhou *et al.* also used hydrothermally synthesized $Ce(OH)_3$ nanotubes as intermediates to prepare CeO_2 nanotubes [70]. The intermediates were exposed in air and then treated with H_2O_2 under sonication.

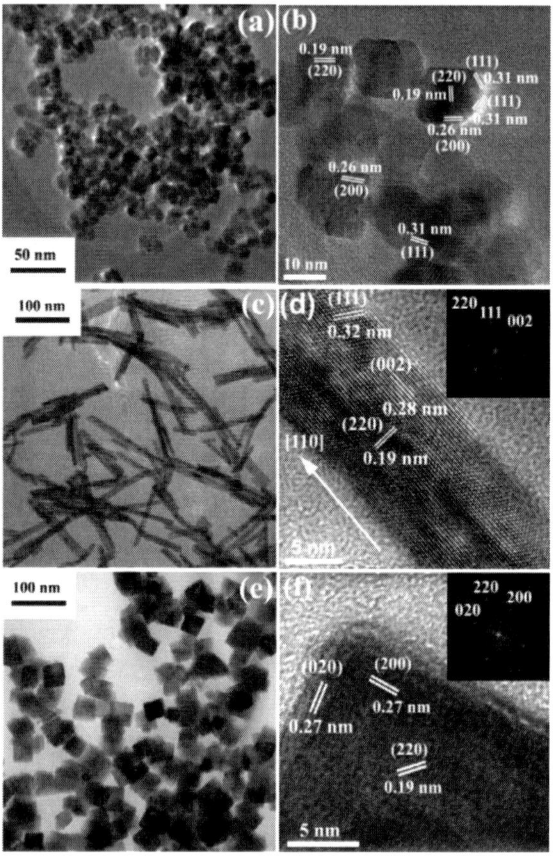

Figure 1.7 TEM (a) and HRTEM (b) images of CeO_2 nanopolyhedra. TEM (c) and HRTEM (d) images of CeO_2 nanorods; inset is a fast Fourier transform (FFT) analysis. TEM (e) and HRTEM (f) images of CeO_2 nanocubes; inset is a FFT analysis.

Nanoceria can also be directly prepared in appropriate nonaqueous or mixed solvents, for example, ethanol [71, 72], CCl_4, and ethanol–water mixed solvents [73]. In a number of cases, addictives like EDTA or fatty acids are found to play crucial roles in controlling the crystallinity and morphology of the products. Sun *et al.* reported that with the assistance of ethylenediamine, phase-pure ceria nanorods (40–50 nm in diameter and 0.3–2 μm in length) were obtained in mixed solvents of ethanol and water [74]. Yang *et al.* prepared high quality ceria nanocubes with strong self-assembly tendency in a water–toluene system, with oleic acid as the stabilizing agent and *tert*-butylamine as the mineralizer (Fig. 1.8) [75]. Li reported the multi-gram one-pot synthesis of monodisperse nanoceria and their superlattices, in mixed solvents of ethanol and water, and with the assistance of oleic acid [76].

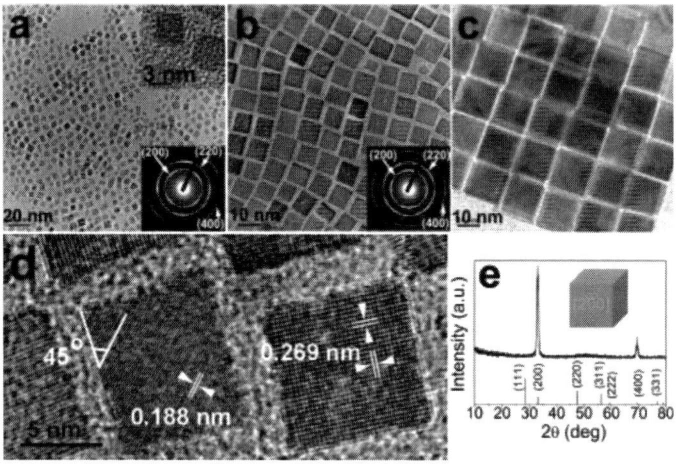

Figure 1.8 (a–c) TEM images of ceria nanocubes with the average sizes of (a) 4.43 nm, (b) 7.76 nm, and (c) 15.65 nm; the insets are SAED patterns and individual nanoparticles. (d) HRTEM image of the ceria nanocubes. (e) A typical XRD pattern of the ceria nanocubes assembled on a Si wafer; the inset is the schematic illustration of the facets of an individual cube.

Cabanas *et al.* developed a continuous hydrothermal route to synthesize $Ce_{1-x}Zr_xO_2$ in critical water flow reactor [77, 78]. Gram-scale products can be prepared at ca. 300°C and 25 MPa. Wright *et al.* reported a one-step hydrothermal synthesis of Ceria–zirconia oxides and found that inclusion of sodium into the crystal structure would lower the reduction temperature of the solids [79].

As for rare earth sesquioxides, the hydrothermal/solvothermal route can afford intermediates (like hydroxides) with uniform shape and size, which can then be converted to corresponding sesquioxides. Li *et al.* reported a series of nanowires, nanotubes, and fullerene-like nanoparticles of rare earth hydroxides and fluorides prepared based on a facile hydrothermal route [80]. The subsequent treatment, such as dehydration, sulfidation, fluoridation, would lead to the formation of nanosized rare earth hydroxides, oxysulfides, and oxyhalides. Similarly, Fang *et al.* also fabricated Y_2O_3, Dy_2O_3, Tb_2O_3 nanotubes from hydrothermally synthesized hydroxide [81, 82]. Hu *et al.* synthesized $La(OH)_3$ nanobelts via a simple hydrothermal route and obtained La_2O_3 nanobelts by post annealing at 300–700°C [83].

Interestingly, Devaraju *et al.* reported the solvothermal synthesis of Y_2O_3:Tb hollow microspheres without any template [84]. Hollow $Y(OH)_3$ and $Y(OH)CO_3$ spheres were obtained first and then converted to the target product. The hollow structure was presumed to be formed under supercritical conditions and due to the Ostwald ripening process. Similar solvothermal-post-annealing synthetic routes were also reported in Y_2O_3:Eu microspheres and hollow microspheres [85, 86]. Chen *et al.* developed a hydrothermal route to synthesize a series of macroporous lanthanide-organic coordination polymer foams, which underwent post-annealing to form corresponding macroporous rare earth oxide monoliths [87]. They also reported rare earth oxide microspheres and hollow spheres prepared by the thermolysis of lanthanide coordination compounds [88].

Li *et al.* performed a series of research works on the hydrothermal synthesis of fluoride nanoparticles of the whole rare earth series, where oleic acid and corresponding sodium salt (or linoleic acid and sodium salt) were used as capping reagents [89, 90]. Lin *et al.* demonstrated a general hydrothermal synthesis method for REF_3 nanocrystals with diverse morphologies (elongated particles, aggregates, and octahedra), where the rare earth nitrates or chlorides, $NaBF_4$, and trisodium citrate were the precursors [91]. They also reported CeF_3, CeF_3:Tb^{3+}, and core–shell structured CeF_3:Tb^{3+}@LaF_3 nanoplates prepared via organic additive and trisodium citrate (Cit^{3-}) assisted hydrothermal method [92]. Chen *et al.* obtained the multicolor upconversion luminescent LaF_3 nanocrystals co-doped with Yb^{3+}–Er^{3+}, Yb^{3+}–Tm^{3+}, and Yb^{3+}–Ho^{3+} ions via a solvothermal route by the decomposition of corresponding rare earth trifluoroacetates [93]. Kumar *et al.* used a similar solvothermal method and synthesized LaF_3:Nd^{3+} nanocrystals exhibiting infrared emissions with high efficiencies [94]. Hollow CeF_3 nanostructures were obtained by Hu *et al.* with the morphologies of nanocages, nanorings, nanococoons, and circular hollow disks [95]. The formation mechanism of hollow nanostructures was also discussed.

Qian *et al.* obtained spindle-like lanthanide-doped YF_3 luminescent nanocrystals [96]. The hydrothermal treatment of YCl_3, NaF, and EDTA at a relatively low temperature of 140°C rendered the products with good uniformity. Yao *et al.* prepared highly uniform and monodisperse truncated octahedral YF_3 nanocrystals in large quantities [97, 98]. Significant activity of Eu^{2+} was observed in the aqueous solution at low temperature (ca. 100°C), and so the Eu^{2+} emission was observed in the Eu-doped YF_3 products. Zeng *et al.* synthesized YOF nanoparticles by the decomposition of a single-source precursor, namely, sodium yttrium fluorocarbonate, under mild hydrothermal conditions, followed by thermal treatments [99].

Li *et al.* reported the hydrothermal synthesis of $NaYF_4:R^{3+}$ (R =Eu, Tb, Yb/Er, and Yb/Tm) submicroprism crystals in hexagonal phase with good uniformity [100]. $RECl_3$, sodium citrate, and NaF were used as starting materials, and the ratio of sodium citrate and RE^{3+} was found to have strong influence on the shape and size of the products (Fig. 1.9). They also synthesized $NaYbF_4$ submicrocrystals with different phase and morphologies, including β-$NaYbF_4$ microdisks, microprisms, microtubes, β-$NaLuF_4$ microprisms, microdisks, microtubes, β-$NaLuF_4$ sub-microplates, α-$NaYbF_4$ submicrospheres, and α-$NaLuF_4$ submicrospheres [101–103]. Wu *et al.* employed an EDTA modified hydrothermal method to prepare prismatic $NaHoF_4$ microtubes and $NaSmF_4$ nanotubes [104].

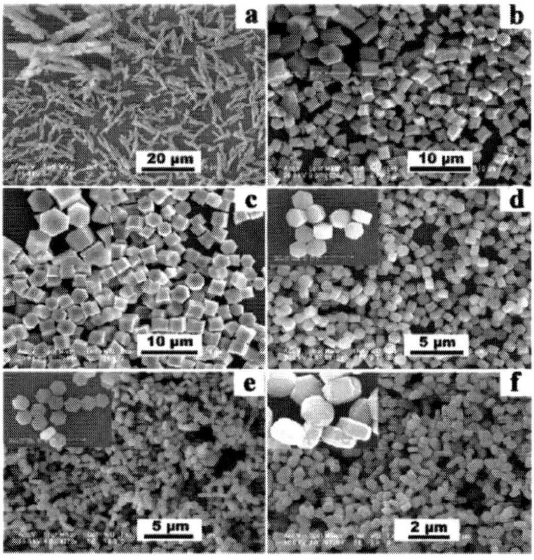

Figure 1.9 SEM images for β-$NaYF_4$:5%Tb^{3+} samples prepared with different molar ratios of sodium citrate and RE^{3+}: (a) without sodium citrate, (b) 1:2, (c) 1:1, (d) 2:1, (e) 4:1, (f) 8:1. Insets are higher-magnification images for the corresponding samples.

The water/alcohol/surfactant solvothermal method developed by Li *et al.* can also be applied to prepare $NaREF_4$ nanocrystals with predictable size, shape, and phase structure, by using $RE(NO_3)_3$, NaF, NaOH, C_2H_5OH, and oleic acid as the starting materials [105–108]. The NaF concentration, rare earth concentration, synthesis temperature, and time govern the size, shape, and phase of $NaREF_4$ nanocrystals. With the assistance of other surfactants such as CTAB and EDTA, more novel $NaYF_4$ nanostructures could be prepared (Fig. 1.10).

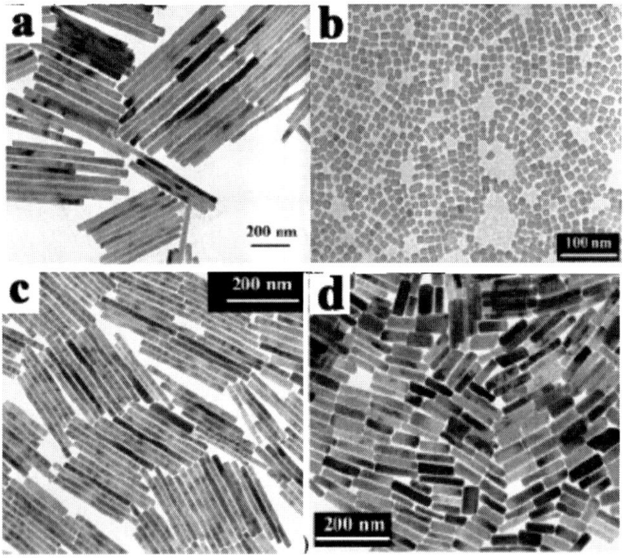

Figure 1.10 TEM image of (a) $NaYF_4:Eu^{3+}$ nanorods, (b) $NaYF_4:Tb^{3+}$ nanocrystals, (c) $NaYF_4:Tb^{3+}$ nanorods, and (d) $NaYF_4:Yb^{3+},Er^{3+}$ nanorods.

Zhao *et al.* adopted the solvothermal method to prepare $NaYF_4$ nanostructured arrays using NaF, $R(NO_3)_3$, and oleic acid [109]. A reverse micelle dissolution–reconstruction process was proposed to interpret the chemical mechanism. They also fabricated uniform β-$NaRF_4$ nanotubes through a hydrothermal in situ ion-exchange reaction by rare earth hydroxide precursors (Fig. 1.11) [110].

Haase *et al.* prepared nanoparticles and nanofibers of $LaPO_4$ with Eu^{3+}, Ce^{3+}, and Tb^{3+} dopant ions [111]. The hydrothermal reaction was carried out at a temperature of 200°C with basic solution of pH 10–12. Similar approach can also give YVO_4 nanoparticles doped with Eu, Sm, and Dy [112, 113]. Yan *et al.* reported the general synthesis of crystalline nanowires of hexagonal and orthorhombic rare earth phosphate hydrate, where RE = La, Ce, Pr, Nd, Sm, Eu, Gd, Tb, Dy [114, 115]. The correlation of radii of rare earth cations and the phase of the final products was discussed. Li *et al.* also reported the controlled synthesis of rare earth phosphate 1D nanomaterials under hydrothermal conditions and the phase, anisotropic growth, and optical properties were discussed [116].

Figure 1.11 (a, b) SEM images of arrays of hexagonal nanotubes of β–$NaYF_4$. (c) TEM image of the nanotubes and corresponding ED patterns (insets). (d) HRTEM image of a nanotube.

Using Pluronic P123 ($EO_{20}PO_{70}EO_{20}$) as surfactant in hydrothermal synthesis, Bu *et al.* obtained single-crystalline $CePO_4$:Tb thin nanorods of typically 10–12 nm in diameter [117]. This surfactant was found to effectively enhance the luminescence emission and chemical homogeneity. At a lower pH value below 1.0, they also obtained uniform spindle like $LaPO_4$ nanowire bundles [118]. Analogously, Guan *et al.* reported the preparation of hollow and core–shell microspheres composed of single-crystalline $CePO_4$ and $CePO_4$:Tb nanorods fabricated by a P123 and $H_6P_4O_{13}$ assisted route [119]. Li *et al.* reported an OA-assisted solvothermal routes in mixed solution of water and ethanol to systematically prepare uniform hexagonal $REPO_4 \cdot nH_2O$ nanocrystals at 140°C (Fig. 1.12) [120]. Under optimized conditions, 0D (sphere-like), 1D (rod-like), and 2D (polygon-like) structures were obtained with good uniformity.

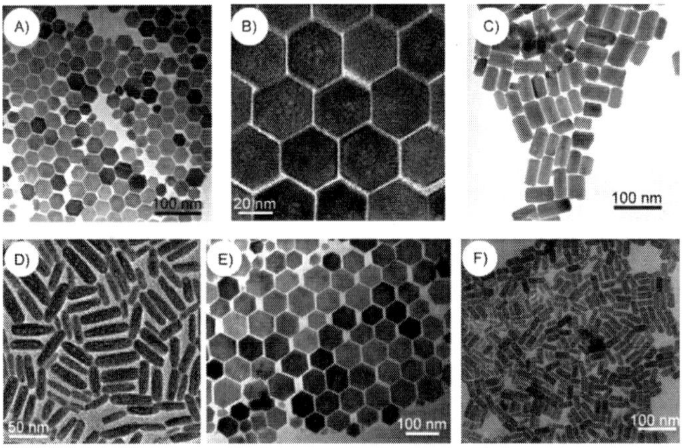

Figure 1.12 TEM image of $REPO_4 \cdot nH_2O$ nanocrystals: (a) Dy, (b) Er, (c) Ho, (d) Tm, (e) Yb, and (f) Lu.

Li reported the hydrothermal synthesis of YVO_4 and YVO_4:Eu nanobelts and microcyrstals under acidic conditions (Fig. 1.13) [121]. The anisotropic growth of nanobelts was found to be influenced by the pH value. The crystal growth along the [010] direction seemed to be less perturbed than the other directions by hydrated proton at pH value of 1. And at higher pH values, microcrystals were obtained.

Figure 1.13 SEM image of YVO_4 nanobelts at different magnifications.

Yan *et al.* employed EDTA-assisted hydrothermal method to prepare *t*-$LaVO_4$:Eu^{3+} nanorods [122, 123]. The metastable zircon structure was found to dramatically enhance the luminescence intensity of $LaVO_4$:Eu^{3+} with respect to the monazite structure prepared following a similar synthetic route. The coordination reagents used play a vital role in the phase transformation process, where weak chelating ligands (acetates and citrates) exert little effect on the formation of tetragonal phase, and strong chelating ligand (EDTA) evidently facilitates the formation of *t*-$LaVO_4$.

The hydrothermal route also show advantages in the preparation of other rare earth oxysalts, particularly titanate, zirconate, and stannate pyrochlores. These materials can be obtained at lower temperatures using hydrothermal reactions, with homogeneous chemical composition, narrow particle size distribution and fine crystallinity as well. Representatively, $La_2Zr_2O_7$ nanoparticles, $La_2Ti_2O_7$ nanosheets, and $La_2Sn_2O_7$ nanospheres were prepared by Chen *et al.*, Li *et al.*, and Moon, respectively [124–126].

1.5 Thermal Decomposition Route

The thermal decomposition synthesis generally resorts to high boiling point solvents, such as trioctylphosphine oxide (TOPO), oleic acid (OA), oleylamine (OM), 1-octadecene (ODE), polyols, and so on. As the reaction proceeds in solution phase, the decomposition of precursors, nucleation, and growth of products can be elaborately controlled either by exquisite adjustment of the internal parameters such as temperature and concentration, or in combination with external injection of other reactants. Consequently, the afforded products usually exhibit good crystallinity, fine size monodispersity, and uniform morphology.

Feldmann reported the polyol-mediated synthesis of a series of functional nanomaterials, including CeO_2, Y_2O_3:Eu, $LaPO_4$:Ce,Tb, and some transition metal oxides and sulfides [127, 128]. The yielded nanoparticles all take a spherical shape, with good crystallinity. Yu *et al.* also employed the polyol method to fabricate uniform ceria nanospheres, microrods, and spindle-like particles [129].

Yan *et al.* synthesized ceria nanoflowers by the pyrolysis of $(NH_4)_2Ce(NO_3)_6$ in OA–OM solvents. The small ceria nanoparticles were formed and then assembled and fused mainly by oriented attachment via [111] faces. They also developed a monitoring method, namely, in situ electrical resistance measurement, which tracks the variation of solution resistance with respect to temperature. The results show that the conductive species are diminished when the flower-like nanostructures form (Fig. 1.14) [130]. Wang *et al.* also reported nanoceria (and other metal oxides) synthesized by the direct thermal decomposition of metal nitrates in OM [131]. Ahniyaz *et al.* prepared ceria nanorods in OA at 200°C [132]. They found that the aspect ratio of the nanorods could be tuned by variation of the reaction time; a long reaction time tends to shorten the nanorods while does not change their diameters. They also found that when OM was introduced, ceria nanodumbbells could be obtained.

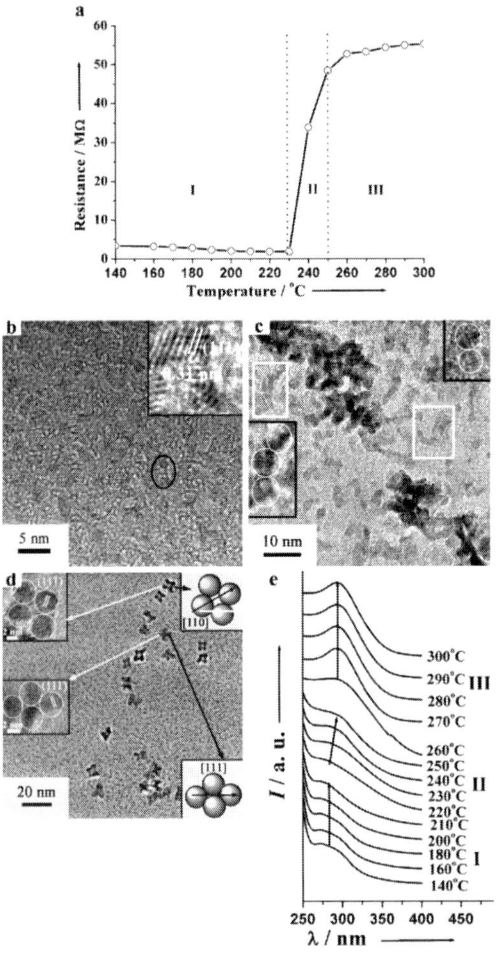

Figure 1.14 (a) Electrical resistance as a function of temperature in the synthesis of CeO_2 nanoflowers. TEM, HRTEM images (inset), and models (inset) of CeO_2 nanocrystals obtained at (b) 140, (c) 230, and (d) 240°C. (e) Dependence of UV/visible absorption on temperature for the synthesis of CeO_2 nanoflowers.

Gu *et al.* reported nanoceria prepared by thermal decomposition of cerium-oleate complex in a number of high boiling point hydrocarbon solvents, including octyl ether, 1-tetradecene, decalin, dipropylene glycol monomethyl ether, dipropylene glycol *n*-butyl ether, and 2,2,4-trimethyl-1,3-pentanediol monoisobutyrate [133]. Hyeon *et al.* developed a large-scale nonhydrolytic sol-gel synthetic route to prepare ceria nanocrystal with spherical, wire-like, and tadpole-like morphologies (Fig. 1.15) [134]. Cerium(III) nitrate was added to different solvents (OM, or mixture of OM/tri-*n*-octylamine, or OM/OA), and diphenyl ether was also introduced. The resulting mixture was heated to 320°C for different durations, and nanospheres, nanowires, and nanotadpoles were obtained.

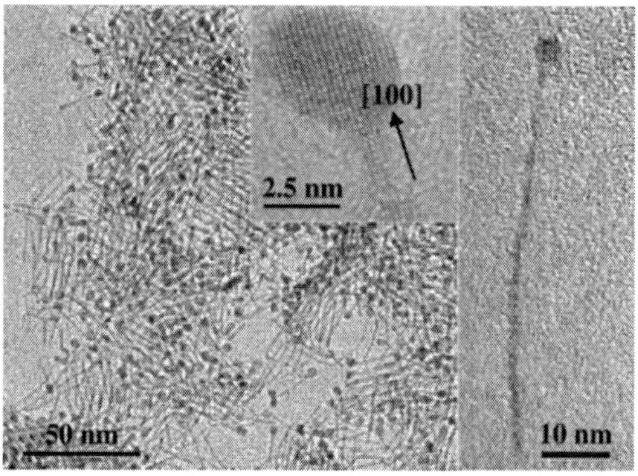

Figure 1.15 TEM images of the tadpole-shaped ceria nanocrystals. The inset is the HRTEM image of a tadpole-shaped nanocrystal.

As for RE_2O_3 nanomaterials, Bazzi *et al.* reported sub-5-nm rare earth oxide nanoparticles prepared in polyalcohol solvents [135, 136]. The nonaqueous environment and the high reaction temperature guaranteed the formation of rare earth oxides, rather than hydroxides. Pinna *et al.* performed a series of research works based on yttria nanostructures [137–140]. The reactions were carried out at in organic solvents at 250–300°C for two days in water-free and oxygen-free environment, and the yielded powders underwent further calcination to form target products.

And a number of RE_2O_3 nanomaterials have been prepared in OA/OM/ODE-mixed solvents via the pyrolysis method. Usually, these obtained nanoparticles exhibit quite uniform morphologies, rather hydrophobic surfaces, and thus strong self-assembly tendencies to form 1D, 2D, and even 3D superstructures. Cao *et al.* used gadolinium acetate as the precursor and obtained square gadolinium oxide nanoplates in OA/OM/ODE solvents at 320°C [141]. Yan *et al.* presented a systematic research work on the pyrolysis synthesis of RE_2O_3 and CeO_2 nanoparticles (Fig. 1.16) [142, 143]. They used different rare earth complexes (benzoylacetonates, acetylacetonates, and acetates) as precursors and obtained nanopolyhedra, nanoplates, and nanodisks with rather narrow size distribution and good crystallinity. These nanomaterials can be readily dispersed in apolar solvents (cyclohexane, toluene, and chloroform, for instance) and can self-organize into fine superlattices after EISA treatment. Heyon *et al.* reported Sm_2O_3 nanowire and nanoplates, where the width of nanowires is of single unit cell thickness (1.1 nm) [144]. Several other RE_2O_3 nanodisks, nanorods, and nanobelts prepared following similar routes were also reported [145–147]. Interestingly, Paek *et al.* obtained Gd_2O_3 nanorings and nanoplates via a two-step process [148]. First, gadolinium acetylacetonate in solvents was hydrolyzed at lower temperature (90°C) in the presence of hydrazine monohydrate. Then the system was heated up to 320°C for thermal dehydration, yielding the nanorings.

Compared to those on rare earth oxides, synthetic research works on rare earth sulfides are relatively less reported, probably due to the fact that trivalent rare earth ions exhibit higher affinity to oxygen rather than sulfur. A major part of related research works adopt dry methods to prepare rare earth (poly)sulfides, such as Y_2S_3, CeS_2, NdS_2, SmS_2, EuS_2, TbS_2, and ErS_2 nanoparticles [149]. EuS, as one of the most extensively studied rare earth sulfides, attracted considerable attention due to its magnetic semiconductor nature. EuS nanoparticles, besides physical routes [150], can also be prepared via chemical routes, among which the thermal decomposition route shows robust viability. Scholes *et al.* reported the synthesis and optical properties of colloidal EuS nanoparticles (Fig. 1.17) [151, 152]. They employed Eu(III) complexes with sulfur-containing ligands, which underwent reduction and decomposition processes in reductive high boiling point solvents and gave EuS nanoparticles. Gao *et al.* also reported the pyrolysis of single-source precursor of a sulfur-containing Eu complex, which was dissolved in oleylamine and heated up to 200–280°C to afford monodisperse EuS nanoparticles [153, 154]. Similar results were also published by Stoll's group [155, 156]. And interestingly, Hasegawa *et al.* employed the photolysis route instead of pyrolysis route to prepare EuS nanoparticles, also with sulfur-containing Eu complexes as the single-source precursor [157].

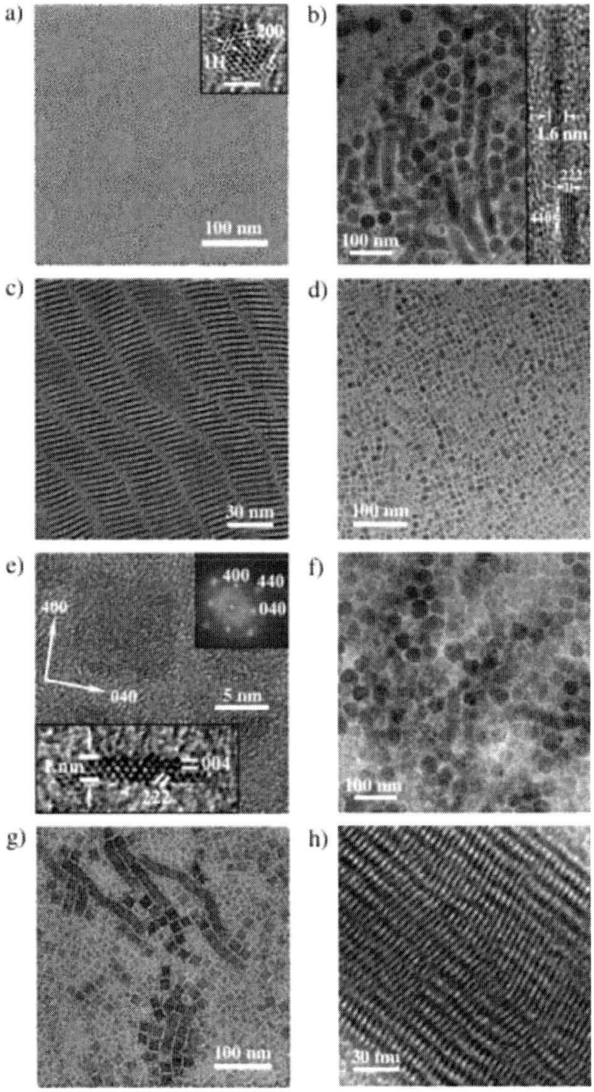

Figure 1.16 TEM image of CeO_2 nanopolyhedra (a), Eu_2O_3 nanodisks and nanoplates and self-assembled superstructures (b, c, d, e), Er_2O_3 nanoplates (f), Pr_2O_3 nanoplates, and self-assembled nanoarrays (g, h).

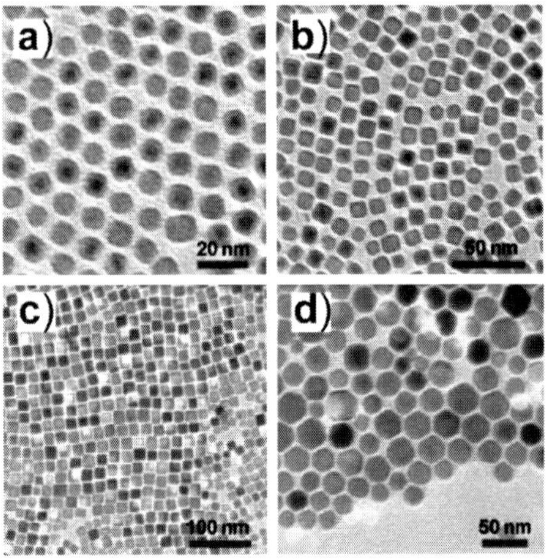

Figure 1.17 TEM images of the four samples of EuS nanoparticles with different morphologies and average diameters.

In contrast to sulfides, rare earth oxysulfides can be obtained under milder conditions. Besides those routes that resort to the sulfurization of intermediates like rare earth hydroxides [158–161], pyrolysis route can also afford high quality rare earth oxysulfides. Gao *et al.* modified the synthetic route such that they prepared rare earth sulfides [162] by introducing oxygen into the reaction system and obtained monodisperse RE_2O_2S nanoparticles from single-source precursors [163].

Lin *et al.* reported colloidal CeF_3, $CeF_3:Tb^{3+}$, and $CeF_3:Tb^{3+}@LaF_3$ nanoparticles yield using the polyol method [164]. Rare earth nitrates and NH_4F were dissolved in diethylene glycol to afford the nanocrystalline products, which exhibit good solubility in water. Song *et al.* employed a similar polyol-mediated route to prepare hexagonal LaF_3 nanoplates in ethylene glycol solvent [165].

Yan *et al.* performed a systematic research works using the pyrolysis of rare earth trifluoroacetates in high boiling point solvents (OA, OM, and ODE) to prepare monodisperse fluorides and oxyfluorides of all rare earth elements (including Y and Sc) [166–168]. The trifluoroacetate ions would release fluoride anions in a controlled manner, and products with good crystallinity, narrow size distribution could be obtained. With aliphatic carbon chains attached on the surface, these nanoparticles can be easily dispersed in apolar solvents and can further self-organize into finely periodic superlattices with EISA technique (Fig. 1.18). This method can also be extended to prepare rare earth oxychlorides, where rare earth trichloroacetates were used as precursors [169]. Similarly, Zhuravleva *et al.* synthesized EuF_3 nanoparticles by the thermal decomposition of $Eu(CF_3COO)_3$ complex in TOPO solvent [170].

By co-thermolysis of CF_3COONa and $RE(CF_3COO)_3$ in organic solvents, Yan *et al.* also obtained high quality $NaREF_4$ nanoparticles with fine uniformity. By the elaborate adjustment of the ratio of Na/RE, solvent composition, reaction temperature, and time, the phase structure and morphology of the $NaREF_4$ nanocrystals could be manipulated. Pure α-$NaREF_4$ could be obtained at a low temperature in OA/OM/ODE solvents within 30 min, while β–$NaRF_4$ was formed at higher temperatures in mixed solvents of OA and ODE (Fig. 1.19) [171]. This method can also be extended to synthesize $LiREF_4$ and $KREF_4$ nanoparticles by using CF_3COOLi and CF_3COOK instead [172].

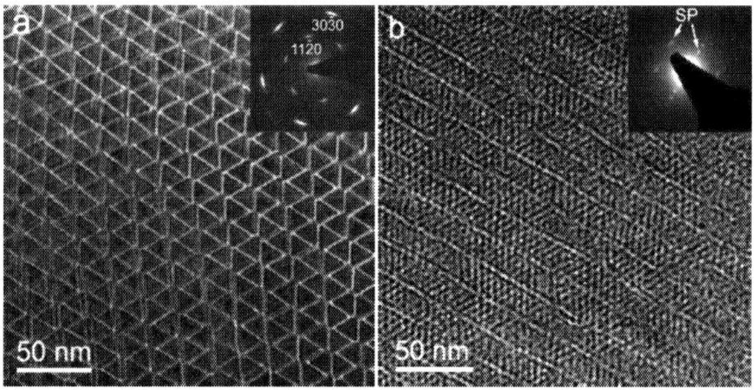

Figure 1.18 TEM images of the (a) edge-to-edge and (b) face-to-face superlattices of LaF_3 nanoplates. Insets show the SAED patterns.

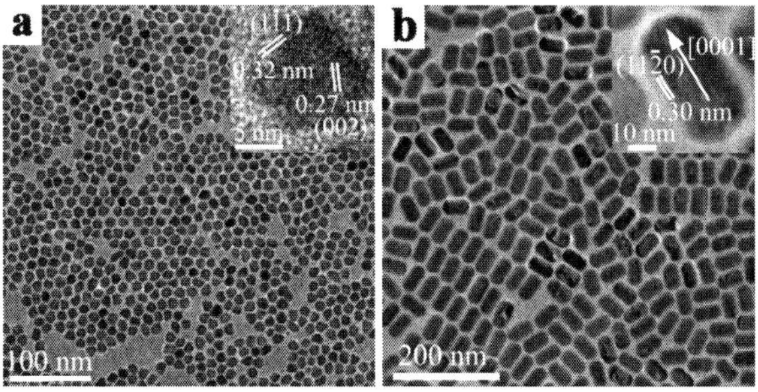

Figure 1.19 TEM and HRTEM (inset) images of high quality (a) cubic and (b) hexagonal phases of $NaYF_4$ nanopolyhedra and nanorods.

Hasse *et al.* presented a set of systematic works on the synthesis of rare earth phosphates and related materials, including $LaPO_4$:Eu, $CePO_4$:Tb, $LuPO_4$:Yb,Tm, and $YbPO_4$:Er, in high-boiling coordinating solvents [173–176]. The obtained nanoparticles can be well redispersed in apolar solvents and can be subjected to a secondary synthetic process to form core–shell nanostructures [177]. Yan *et al.* exploited the OA/OM/ODE mixture solvents to synthesize a series of high quality dispersible $REPO_4$ nanocrystals with various shapes, including nanopolyhedra, quasinanorods, nanorods, and nanowires. The reaction was carried out at temperatures in the range of 180–260°C in oleic acid and oleylamine solvents, and the products were formed via a limited anion-exchange mechanism (Fig. 1.20) [178].

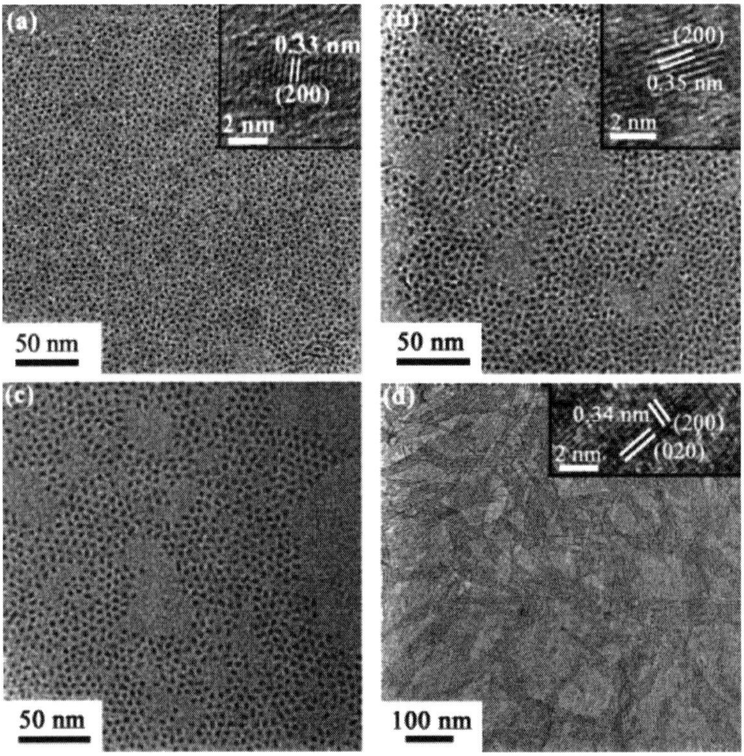

Figure 1.20 TEM and HRTEM images (insets) of (a) $LaPO_4$ polyhedra and (b) $EuPO_4$ polyhedra. (c) TEM image of $TbPO_4$ polyhedra. (d) TEM and HRTEM images (inset) of YPO_4 nanowires.

1.6 Microwave Route

The microwave route is usually featured by its short reaction time, facile operation, and robust productivity and is sometimes employed in combination with other synthetic systems (hydrothermal autoclaves, for example). Different from other heating mechanisms, the energy of microwaves is almost simultaneously applied to the whole reaction system, thus obviating the thermal gradient effects in other synthetic routes. Also due to the high penetrability of microwaves, the local temperature at the microscopic reaction sites can be abruptly elevated to a sufficiently high value, while the overall temperature of the reaction system is considerably lower. Therefore, the rapid decomposition of precursors and the explosive nucleation of the products are facilitated, resulting in finely sized nanoparticles within drastically decreased temporal durations.

Liao *et al.* reported the preparation of monodisperse nanocrystalline CeO_2 powders by microwave irradiation [179]. Ultrafine crystalline ceria powder of ca. 2.0 nm were obtained by hydrolysis of $(NH_4)_2Ce(NO_3)_6$ in aqueous solution containing PEG-2000 and NaAc.

Leite *et al.* utilized microwave as the heating source of the hydrothermal synthesis of Gd-dope ceria nanorods [180]. The anisotropic growth was mainly ascribed to the oriented attachment accelerated by effective collision of the nanoparticles. Microwave irradiation was believed to increase the collision cross section and thus increase the effective collision rate. The duration of hydrothermal treatment was also largely decreased.

Panda *et al.* combined the microwave technique with high boiling point solvent pyrolysis route and obtained a series of rare earth sesquioxides in OA and OM within several minutes [181]. By varying the reaction time and concentration of organic surfactants, the size and morphology of the yielded nanostructures can be controlled. Similarly, Hasegawa *et al.* also employed the microwave technique to promote the thermal decomposition of Eu complexes in acetonitrile to give EuS nanoparticles [182].

Ma *et al.* employed the microwave-assisted hydrothermal method to obtain hollow-structured PrF_3 nanoparticles, and they proposed that the dissolution–recrystallization process was probably responsible for the formation of the hollow structure [183].

1.7 Microemulsion Route

The microemulsion method utilizes a water/oil/surfactant system to construct numerous micro- and nanometer reactors, in which the nanoparticles are formed. Macroscopically, microemulsion is an optically isotropic and thermodynamically stable disperse system. At nanometer scale, the microemulsion is heterogeneous with an internal structure either of nanospherical droplets (micelles or reverse micelles) or a bicontinuous phase, depending on the given temperature as well as the ratio of different constituents. The droplets are generally of similar size, and each small droplet could be utilized as a micro-/nanoreactor, which promises the nanoparticles to be prepared in a very controllable manner.

Adachi prepared ultrafine ceria nanoparticles within a reverse microemulsion method [184]. Reverse micelles were formed in an aqueous solution of polyoxyethylene(10) octylphenyl ether, n-hexyl alcohol, and cyclohexane. The mean particle size obtained was between 2.6 and 4.1 nm, exhibiting a fairly narrow size distribution. Similar methods were adopted by Bumajad *et al.* and Sathyamurthy *et al.* to prepare nanoceria with high surface areas and good thermal stabilities [185, 186].

Shi *et al.* employed the reverse micelle system to prepare CeF_3:Lu^{3+} nanoparticles, where the emulsion was composed of CTAB, *n*-butanol, *n*-octane, and water. The characteristics emission of Lu^{3+} was observed [187]. Ritcey *et al.* prepared quadrilateral-shaped YF_3 nanoparticles with water in cyclohexane reverse microemulsion under the stabilization of polyoxyethylene isooctylphenyl ether (igepal CO520) [188]. The microemulsion was formed by mixing aqueous YCl_3 and Igepal CO520 in cyclohexane under stirring, then the aqueous solution of NH_4HF_2 was added to produce the YF_3 nanoparticles within every micelle water pool. The as-obtained nanoparticles exhibit highly uniform shapes and can self-assemble into periodic superstructures (Fig. 1.21). Similarly, Wang *et al.* obtained YF_3:Yb,Tm nanobundles via the microemulsion method using water, CTAB, cyclohexane, and 1-pentanol [189].

Cao *et al.* developed a hydrothermal microemulsion method to prepare $LaPO_4$ and $CePO_4$ nanorods/nanowires under mild conditions at 140°C [190]. The aspect ratio of the as-obtained nanostructures was found to be dependent on the water/surfactant molar ratio. Xiang *et al.* reported the synthesis of uniform $CePO_4$ nanorods by the reaction of aqueous micelles containing $(CTA)_3PO_4$ with reverse micelles containing $Ce(AOT)_3$ (CTA = cetyltrimethylammonium and AOT = bis(2-ethylhexyl)sulfosuccinate) prepared in isooctane [191]. Similarly, using the reverse micelles system, Ghosh *et al.* reported the synthesis of $LaPO_4$:Er,Yb and $LaPO_4$:Er@$YbPO_4$ nanoparticles and nanorods and found that the core–shell structured nanostructure helps to enhance the luminescence efficiency [192].

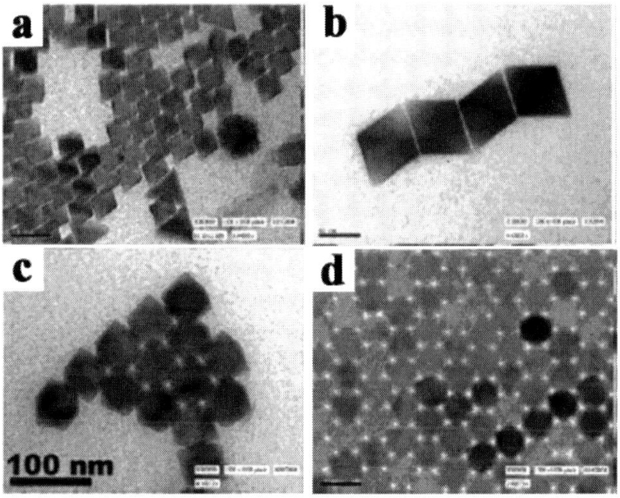

Figure 1.21 TEM images of different morphologies of YF_3 nanocrystals: (a) typical mixtures of predominant YF_3 particles morphologies; (b) YF_3 quadrilateral-shaped nanocrystals; (c) YF_3 hexagonal nanocrystals; (d) self-assembly of YF_3 hexagonal nanocrystals.

1.8 Other Routes

Zhang prepared ceria nanorods via ultrasonication at room temperature with the assistance of polyethylene glycol [193]. The nanorods are 5–10 nm in diameter and 50–150 nm in length. The formation may be ascribed to the oriented aggregation and fusing of small ceria nanoparticles with the aid of ultrasonication. Zhu *et al.* reported the sonochemical synthesis of $CePO_4$:Tb, $CePO_4$:Tb@ $LaPO_4$, and YVO_4:Eu nanorods [194, 195]. The products were yielded within reduced time but still with good crystallinity and uniformity. Yu *et al.* developed a surfactant-free sonochemical route for $REPO_4$ nanocrystals under ambient conditions [196]. The afforded products in hexagonal phase (RE = La, Ce, Pr, Nd. Sm, Eu, Gd) exhibited nanorod bundles morphology, and the products in tetragonal phase (RE = Tb, Dy, Ho) exhibited particle-like morphology.

Yan *et al.* reported the preparation of nearly monodisperse spherical aggregates of CeO_2 nanocrystals in ionic liquids (IL), where the IL 1-hexadecyl-3-methylimidazolium bromide (C16MimBr) acted as both the template and solvent. The spherical aggregates were 100–150 nm in diameter, composed of small CeO_2 nanocrystals (ca. 3.5 nm) as building units, and thus exhibiting 3D open mesoporous structures [197]. They also used the IL route to prepare $NaYF_4$ nanoclusters in combination with microwave irradiation [198]. Chen *et al.* synthesized REF_3 (RE = La, Ce, Pr, Nd, Sm, Eu, and Er), $CeF_3:Tb^{3+}$, as well as $LaF_3:Eu^{3+}$ nanocrystals with good uniformity in different ILs (1-octyl-3-methylimidazolium hexafluorophosphate, 1-octyl-3-methylimidazolium tetrafluoroborate, and 1-butyl-3-methylimidazolium hexafluorophate). The ILs acted as both solvents and templates, and during the reaction process, the PF_6^- and BF_4^- anions underwent partial hydrolysis and thus provided the fluorine source [199]. Bühler and Feldmann reported microwave-assisted synthesis of $LaPO_4$:Ce,Tb nanocrystals in ILs [200]. Typically, a solution of rare earth salts in IL and cosolvent was added to another portion of IL solution containing phosphate precursors, and the obtained opalescent mixture was heated by microwave at 300°C for 10 s.

Okuyama *et al.* employed a modified aerosol decomposition method to prepare CeO_2 particles, with addition of eutectic salts to the solution [201]. The eutectic salts were found to be helpful to reduce the particle size and enhance the crystallinity. Similarly, Hu *et al.* also used eutectic salts in the composite-hydroxide-mediated synthesis of ceria [202]. Huang *et al.* fabricated ultrathin crystalline Gd-dope ceria films by dc sputtering and subsequent oxidation treatment [203].

Li *et al.* reported the general synthesis of a series of metal oxide hollow spheres (SnO_2, Al_2O_3, La_2O_3, Y_2O_3, Lu_2O_3, CeO_2, TiO_2, ZrO_2, etc.) using carbonaceous polysaccharide microspheres as templates (Fig. 1.22) [204]. The metal sources in solution were absorbed onto the functional surface layer of carbonaceous saccharide microspheres, and subsequent calcinations and oxidation allowed the absorbed species to densify and cross-link into hollow spheres, while the templates were removed.

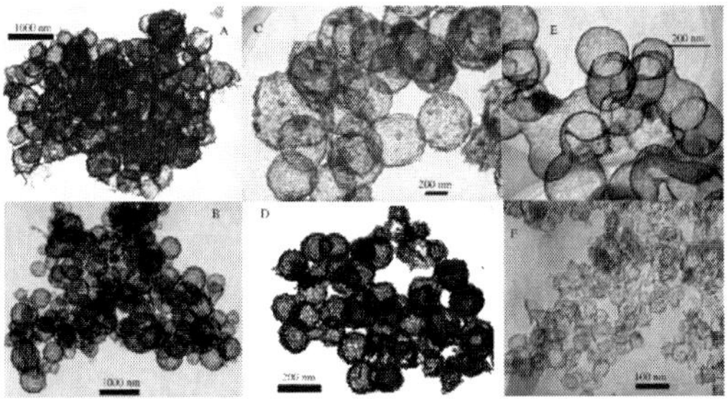

Figure 1.22 TEM images of transition metal and rare earth oxide hollow spheres: (a) CoO, (b) Mn_3O_4, (c) NiO, (d) Cr_2O_3, (e) CeO_2, (f) Lu_2O_3.

Karakoti *et al.* reported the self-assembly of small ceria nanoparticles into nanorods during the structural formation of ice (Fig. 1.23) [205]. They proposed that during the freezing and the aging (2–3 weeks) of the ceria suspension, the trapped nanoparticles tend to combine via an oriented attachment process, which results in the formation of nanorods.

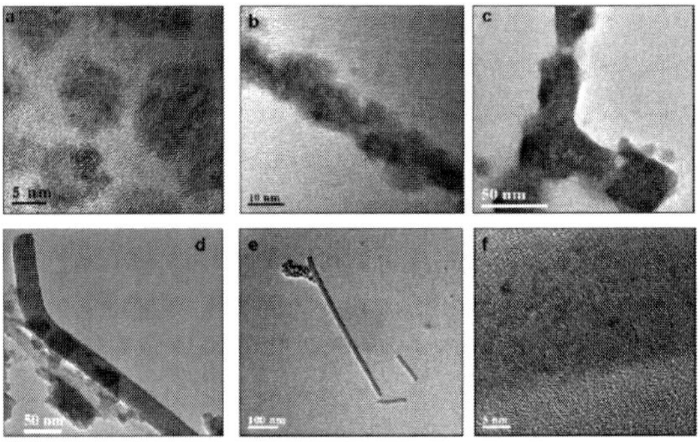

Figure 1.23 TEM images of the ceria solution: (a) immediately after the formation of ceria nanoparticles (10–15 nm agglomerates contains 3–5 nm ceria nanocrystals), and after freezing and subsequent aging for (b) one day—the zigzag alignment of nanoparticles into a 1D self-assembled structure is apparent; (c) one week—initial ceria nanocrystals agglomerated in ice arrange anisotropically to form long nanorods; (d) samples aged for one week showing complete nanorods; (e) two weeks—a long nanorod (aspect ratio 1:20) formed after aging in ice; (f) two weeks—high magnification image of the nanorod in (e) showing the polycrystalline nature of nanorods with 3–5 nm nanocrystallites.

Yanagida *et al.* reported EuS nanoparticles prepared in liquid ammonia [206]. The typical synthetic procedures involve the dissolution of Eu metal in liquid ammonia and the introduction of H_2S as the sulfur source. The yielded nanoparticles can be collected by simply evaporating the solvent. A similar route reported by Kataoka *et al.* used thiourea as the sulfurization reagent [207].

1.9 Summary

The study on the synthesis of rare earth nanomaterials is still ongoing due to numerous interests of such materials in various applications. Though considerable success has been achieved up to date, several challenging scientific aspects of various synthesis techniques are yet to be addressed:

1. The detailed formation mechanisms of nanomaterials in many systems are yet unclear, or simply in the stage of presumptions, which lacks concrete and supportive experimental evidences. Thus, it is of urgent necessity to develop in situ and real-time monitoring methodologies with high sensitivities, so as to allow for insightful perspectives of the reaction process.
2. The synthetic routes developed so far seem less effective for some systems, for example, Y-Ba-Cu-O-type compounds, Eu(II)-associated materials, metallic rare earth nanomaterials, owing to either the complexity in chemical composition and reactivity of each individual species, or the stringent experimental conditions required.
3. Nanoparticles with good crystallinity, few defects, and high applicability are still in urgent demand, especially those with luminescent properties, in which case the defects present in the nanoparticles would severely quench the excited emissive species, and thus largely hamper the luminescent efficiencies.
4. The fabrication methods for hybrid/composite nanostructures, which could afford integrated functionalities and improved performances, are yet limited and usually involve laborious fabrication procedures. Simple and efficient methods are still desired to assemble the rare earth nanomaterials with other functional entities, such as organic dyes, semiconductor quantum dots, noble metals, and magnetic nanoparticles.

Reference

1. Zhang, F.; Chan, S.W.; Spanier, J.E.; Apak, E.; Jin, Q.; Robinson, R.D.; Herman, I.P. (2002) *Appl. Phys. Lett.*, **80**, 127.
2. Han, W.Q.; Wu, L.J.; Zhu, Y.M.J. (2005) *Am. Chem. Soc.*, **127**, 12814.
3. Du, N.; Zhang, H.; Chen, B.G.; Ma, X.Y.; Yang, D.R. (2007) *J. Phys. Chem. C*, **111**, 12677.
4. Deshpande, A.S.; Pinna, N.; Beato, P.; Antonietti, M.; Niederberger, M. (2004) *Chem. Mater.*, **16**, 2599.
5. Deshpande, A.S.; Niederberger, M. (2007) *Microporous Mesoporous Mater.*, **101**, 413.
6. Wakefield, G.; Keron, H.A.; Dobson P.J.; Hutchison, J.L. (1999) *J. Colloid. Interface Sci.*, **215**, 179.
7. Wakefield, G.; Holland, E.; Dobson, P.J.; Hutchison, J.L. (2001) *Adv. Mater.*, **13**, 1557.
8. Yada, M.; Mihara, M.; Mouri, S.; Kuroki, M.; Kijima, T. (2002) *Adv. Mater.*, **14**, 309.
9. Yada, M.; Taniguchi, C.; Torikai, T.; Watari, T.; Furuta, S.; Katsuki, H. (2004) *Adv. Mater.*, **16**, 1448.
10. Stouwdam, J.W.; van Veggel, F.C.J.M. (2002) *Nano Lett.*, **2**, 733.
11. Yi, G.S.; Chow, G.M. (2005) *J. Mater. Chem.*, **15**, 4460.
12. Wang, M.; Huang, Q.L.; Hong, J.M.; Chen, X.T.; Xue, Z.L. (2006) *Cryst. Growth Des.*, **6**, 2169.
13. Wang, M.; Huang, Q.L.; Hong, J.M.; Chen, X.T.; Xue, Z.L. (2006) *Cryst. Growth Des.*, **6**, 1972.
14. Zhu, L.; Li, Q.; Liu, X.D.; Li, J.Y.; Zhang, Y.F.; Meng, J.; Cao, X.Q. (2007) *J. Phys. Chem. C*, **111**, 5898.
15. Zhu, L.; Liu, X.M.; Meng, J.; Cao, X.Q. (2007) *Cryst. Growth Des.*, **7**, 2505.
16. Miao, Z.J.; Liu, Z.M.; Ding, K.L.; Han, B.X.; Miao, S.D.; An, G.M. (2007) *Nanotechnology*, **18**, 125605.
17. Zhang, Y.; Lu, M.H. (2007) *Nanotechnology*, **18**, 275603.
18. Ansari, A.A.; Singh, N.S.S.P. (2008) *J. Nanopart. Res.*, **10**, 703.
19. Wang, M.; Huang, Q.L.; Hong, J.M.; Wu, W.H.; Yu, Z.; Chen, X.T.; Xue, Z.L. (2005) *Solid State Commun.*, **136**, 210.
20. Yi, G.S.; Lu, H.C.; Zhao, S.Y.; Ge,Y.; Yang, W.J.; Chen, D.P.; Guo, L.H. (2004) *Nano Lett.*, **4**, 2191.

21. Karbowiak, M.; Bednarkiewicz, A. (2004) *J. Alloy Compd.*, **380**, 321.
22. Karbowiak, M.; Mech, A.; Bednarkiewicz, A.; Strek, W.; Kepinski, L. (2005) *J. Phys. Chem. Solids*, **66**, 1008.
23. Wang, X.J.; Gao, M.Y. (2006) *J. Mater. Chem.*, **16**, 1360.
24. Di, W.H.; Zhao, X.X.; Lu, S.Z.; Wang, X.J.; Zhao, H.F. (2007) *J. Solid State Chem.*, **180**, 2478.
25. Buissette, V.; Moreau, M.; Gacoin, T.; Boilot, J.P.; Chane-Ching, J.Y.; Le Mercier, T. (2004) *Chem. Mater.*, **16**, 3767.
26. Huignard, A.; Gacoin, T.; Boilot, J.P. (2000) *Chem. Mater.*, **12**, 1090.
27. Huignard, A.; Buissette, V.; Laurent, G.; Gacoin, T.; Boilot, J.P. (2002) *Chem. Mater.*, **14**, 2264.
28. Takeshita, S.; Isobe, T.; Niikura, S. (2008) *J. Lumin.*, **128**, 1515.
29. Stouwdam, J.W.; Raudsepp, M.; van Veggel, F.C.J.M. (2005) *Langmuir*, **21**, 7003.
30. Sahu, P.K.; Behera, S.K.; Pratihar, S.K.; Bhattacharyya, S. (2004) *Ceram. Int.*, **30**, 1231.
31. Shlyakhtina, A.V.; Abrantes, J.C.C.; Larina, L.L.; Shcherbakova, L.G. (2005) *Solid State Ionics*, **176**, 1653.
32. Wang, S.; Zhou, G.; Lu, M.; Zhou, Y.; Wang, S.; Yang, Z. (2006) *J. Am. Ceram. Soc.*, **89**, 2956.
33. Marrero-Lopez, D.; Nunez, P.; Abril, M.; Lavin, V.; Rodriguez-Mendoza, U.R.; Rodriguez, V.D. (2004) *J. Non-Cryst. Solids*, **345 and 346**, 377.
34. Marrero-Lopez, D.; Pena-Matinez, J.; Ruiz-Morales, J.C.; Nunez, P. (2008) *J. Solid State Chem.*, **181**, 253.
35. Kharrazi, S.; Kundaliya, D.C.; Gosavi, S.W.; Kulkarni, S.K.; Venkatesan, T.; Ogale, S.B.; Urban, J.; Park, S.; Cheong S.-W. (2006) *Solid State Commun.*, **138**, 395.
36. Liang, S.; Teng, F.; Bulgan, G.; Zhu, Y. (2007) *J. Phys. Chem. C*, **111**, 16742.
37. White, B.; Chatterjee, S.; Macam, E.; Wachsman, E. (2008) *J. Am. Ceram. Soc.*, **91**, 2024.
38. Brezesinski, T.; Erpen, C.; Iimura, K.; Smarsly, B. (2005) *Chem. Mater.*, **17**, 1683.
39. Thammachart, M.; Meeyoo, V.; Risksomboon, T.; Osuwan, S. (2001) *Catal. Today*, **68**, 53.
40. Yuan, Q.; Liu, Q.; Song, W.G.; Feng, W.; Pu, W.L.; Sun, L.D.; Zhang, Y.W.; Yan, C.H. (2007) *J. Am. Chem. Soc.*, **129**, 6698.

41. Goldburt E.T.; Kulkarni B.; Bhargava R.N.; Taylor J.; Libera M. (1997) *J. Lumin.*, **72–74**, 190.

42. Dhanaraj, J.; Jagannathan, R.; Kutty, T.R.N.; Lu, C.H. (2001) *J. Phys. Chem. B*, **105**, 11098.

43. Wu, G.S.; Zhang, L.D.; Cheng, B.C.; Xie, T.; Yuan, X.Y. (2004) *J. Am. Chem. Soc.*, **126**, 5976.

44. Hou, Z.Y.; Yang, P.P.; Li, C.X.; Wang, L.L.; Lian, H.Z.; Quan, Z.W.; Lin, J. (2008) *Chem. Mater.*, **20**, 6686.

45. Hreniak, D.; Strek, W.; Deren, P.; Bednarkiewicz, A.; Lukowiak, A. (2006) *J. Alloy Compd.*, **408–412**, 828.

46. Fujihara, S.; Tokumo, K. (2005) *Chem. Mater.*, **17**, 5587.

47. Li, J.Y.; Dai, H.; Li, Q.; Zhong, X.H.; Ma, X.F.; Meng, J.; Cao, X.Q. (2006) *Mater. Sci. Eng. B*, **133**, 209.

48. Subramania, A.; Saradha, T.; Muzhumathi, S. (2007) *J. Power Sources*, **167**, 319.

49. Popa M.; Frantti J.; Kakinaha M. (2002) *Solid State Ionics*, **154–155**, 135.

50. Popa M.; Frantti J.; Kakinaha M. (2002) *Solid State Ionics*, **154–155**, 437.

51. Popa M.; Hong L.V.; Kakinaha M. (2003) *Phys. B*, **327**, 233.

52. Popa M.; Hong L.V.; Kakinaha M. (2003) *Phys. B*, **327**, 237.

53. Niu, X.S.; Du, W.M.; Du, W.P.; Jiang, K. (2003) *J. Rare Earth*, **21**, 630.

54. Niu, X.S.; Du, W.M.; Du, W.P.; Jiang, K. (2004) *Chin. J. Chem. Phys.*, **17**, 79.

55. Niu, X.S.; Du, W.M.; Du, W.P. (2004) *Sens. Act. B*, **99**, 399.

56. Niu, X.S.; Du, W.M.; Du, W.P.; Jiang, K. (2005) *Rare Met. Mater. Eng.*, **34**, 124.

57. Niu, X.S.; Li, H.H.; Liu, G.G. (2005) *J. Mol. Catal. A Chem.*, **232**, 89.

58. Niu, X.S.; Li, H.H.; Zhang, F.; Liu, G.G.; Jiang, K. (2005) *J. Rare Earth.*, **23**, 420.

59. Li, G.S.; Feng, S.H.; Li, L.P. (1996) *J. Solid State Chem.*, **126**, 74.

60. Lu, C.H.; Wang, H.C. (2002) *Mater. Sci. Eng. B*, **90**, 138.

61. Han, Z.H.; Guo, N.; Tang, K.B.; Yu, S.H.; Zhao, H.Q.; Qian, Y.T. (2000) *J. Cryst. Growth*, **219**, 315.

62. Chen, S.F.; Yu, S.H.; Yu, B.; Ren, L.; Yao, W.T.; Cölfen, H. (2004) *Chem. Eur. J.*, **10**, 3050.

63. Guo, Z.Y.; Du, F.L.; Li, G.C.; Cui, Z.L. (2006) *Inorg. Chem.*, **45**, 4167.

64. Guo, Z.Y.; Du, F.L.; Li, G.C.; Cui, Z.L. (2008) *Cryst. Growth Des.*, **8**, 2674.

65. Sun, C.W.; Sun, J.; Xiao, G.L.; Zhang, H.R.; Qiu, X.P.; Li, H.; Chen, L.Q. (2006) *J. Phys. Chem. B*, **110**, 13445.

66. Chen, G.Z.; Sun, S.X.; Song, X.Y.; Yin, Z.L.; Yu, H.Y.; Fan, C.H. Zhao, W. (2007) *J. Mater. Sci.*, **42**, 6977.

67. Yan, L.; Yu, R.B.; Chen, J.; Xing, X.R. (2008) *Cryst. Growth Des.*, **8**, 1474.

68. Mai, H.X.; Sun, L.D.; Zhang, Y.W.; Si, R.; Feng, W.; Zhang, H.P.; Liu, H.C.; Yan, C.H. (2005) *J. Phys. Chem. B*, **109**, 24380.

69. Tang, C.C.; Bando, Y.; Liu, B.D.; Golberg, D. (2005) *Adv. Mater.*, **17**, 3005.

70. Zhou, K.B.; Yang, Z.Q.; Yang, S. (2007) *Chem. Mater.*, **19**, 1215.

71. Verdon, E.; Devalette, M.; Demazeau, G. (1995) *Mater. Lett.*, **25**, 127.

72. Wang, C.Y.; Qian Y.T.; Xie Y.; Wang C.S.; Yang L.; Zhao G.W. (1996) *Mater. Sci. Eng. B*, **39**, 160.

73. Wang, C.Y.; Zhang, W.Y.; Qian, Y.T. (2002) *Mater. Sci. Eng. B*, **94**, 170.

74. Sun, C.W.; Li, H.; Zhang, H.R.; Wang, Z.X.; Chen, L.Q. (2005) *Nanotechnology*, **16**, 1454.

75. Yang, S.W.; Gao, L. (2006) *J. Am. Chem. Soc.*, **128**, 9330.

76. Huo, Z.Y.; Chen, C.; Liu, X.W.; Chu, D.R.; Li, H.H.; Peng, Q.; Li, Y.D. (2008) *Chem. Commun.*, **32**, 3741.

77. Cabanas, A.; Darr, J.A.; Lester, E.; Poliakoff, M. (2000) *Chem. Commun.*, **11**, 901.

78. Cabanas, A.; Darr, J.A.; Lester, E.; Poliakoff, M. (2001) *J. Mater. Chem.*, **11**, 561.

79. Wright, C.S.; Walton, R.I.; Thompsett, D.; Fisher, J.; Ashbrook, S.E. (2007) *Adv. Mater.*, **19**, 4500.

80. Wang, X.; Li, Y.D. (2003) *Chem. Eur. J.*, **9**, 5627.

81. Fang, Y.P.; Xu, A.W.; Song, R.Q.; Zhang, H.X.; You, L.P.; Yu, J.C.; Liu, H.Q. (2003) *J. Am. Chem. Soc.*, **125**, 16025.

82. Fang, Y.P.; Xu, A.W.; You, L.P.; Song, R.Q.; Yu, J.C.; Zhang, H.X.; Li, Q.; Liu, H.Q. (2003) *Adv. Func. Mater.*, **13**, 955.

83. Hu, C.G.; Liu, H.; Dong, W.T.; Zhang, Y.Y.; Bao, G.; Lao, C.S.; Wang, Z.L. (2007) *Adv. Mater.*, **19**, 470.

84. Devaraju, M.K.; Yin, S.; Sato, T. (2009) *Cryst. Growth Des.*, **9**, 2944.

85. Yang, J.; Quan, Z.W.; Kong, D.Y.; Liu, X.M.; Lin, J. (2007) *Cryst. Growth Des.*, **7**, 730.

86. Jia, G.; Yang, M.; Song, Y.H.; You, H.P.; Zhang, H.J. (2009) *Cryst. Growth Des.*, **9**, 301.

87. Shen, Z.R.; Zhang, G.J.; Zhou, H.J.; Sun, P.C.; Li, B.H.; Ding, D.T.; Chen, T.H. (2008) *Adv. Mater.*, **20**, 984.

88. Shen, Z.R.; Wang, J.G.; Sun, P.C.; Ding, D.T.; Chen, T.H. (2009) *Chem. Commun.*, 1742.

89. Wang, X.; Zhuang, J.; Peng, Q.; Li, Y.D. (2006) *Inorg. Chem.*, **45**, 6661.

90. Wang, L.Y.; Li, P.; Li, Y.D. (2007) *Adv. Mater.*, **19**, 3304.

91. Li, C.X.; Yang, J.; Yang, P.P.; Lian, H.Z.; Lin, J. (2008) *Chem. Mater.*, **20**, 4317.

92. Li, C.X.; Liu, X.M.; Yang, P.P.; Zhang, C.M.; Lian, H.Z.; Lin, J. (2008) *J. Phys. Chem. C*, **112**, 2904.

93. Liu, C.H.; Wang, H.; Zhang, X.R.; Chen, D.P. (2009) *J. Mater. Chem.*, **19**, 489.

94. Kumar, G.A.; Chen, C.W.; Ballato, J.; Riman, R.E. (2007) *Chem. Mater.*, **19**, 1523.

95. Wu, Q.; Chen, Y.; Xiao, P.; Zhang, F.; Wang, X.Z.; Hu, Z. (2008) *J. Phys. Chem. C*, **112**, 9604.

96. Zhang, M.F.; Shi, S.J.; Meng, J.X.; Wang, X.Q.; Fan, H.; Zhu, Y.Y.; .Wang, X.Y.; Qian, Y.T. (2008) *J. Phys. Chem. C*, **112**, 2825.

97. Tao, F.; Wang, Z.J.; Yao, L.Z.; Cai, W.L.; Li, X.G. (2007) *Cryst. Growth Des.*, **7**, 854.

98. Tao, F.; Wang, Z.J.; Yao, L.Z.; Cai, W.L.; Li, X.G. (2007) *J. Phys. Chem. C*, **111**, 3241.

99. Zeng, J.H.; Lou, T.J.; Wang, Y.F.; Guo, J.C.; Di, D.; Ma, R.L. (2009) *J. Phys. Chem. C*, **113,** 597.

100. Li, C.X.; Quan, Z.W.; Yang, J.; Yang, P.P.; Lin, J. (2007) *Inorg. Chem.*, **46**, 6329.

101. Li, C.X.; Quan, Z.W.; Yang, P.P.; Huang, S.S.; Lian, H.Z.; Lin, J. (2008) *J. Phys. Chem. C*, **112**, 13395.

102. Li, C.X.; Quan, Z.W.; Yang, P.P.; Yang, J.; Lian, H.Z.; Lin, J. (2008) *J. Mater. Chem.*, **18**, 1353.

103. Li, C. X.; Quan, Z.W.; Yang, P.P.; Zhang, X.M.; Lian, H.Z.; Lin, J. (2008) *Cryst. Growth Des.*, **8**, 923.

104. Liang, L.; Xu, H.F.; Su, Q.; Konishi, H.; Jiang, Y.B.; Wu, M.M.; Wang,Y.F.; Xia, D.Y. (2004) *Inorg. Chem.*, **43**, 1594.

105. Wang, L.Y.; Li, P.; Li, Y.D. (2007) *Adv. Mater.*, **19**, 3304.

106. Liang, X.; Wang, X.; Zhuang, J.; Peng, Q.; Li, Y.D. (2007) *Adv. Funct. Mater.*, **17**, 2757.

107. Wang, L.Y.; Li, Y.D. (2007) *Chem. Mater.*, **19**, 727.

108. Zeng, J.H.; Su, J.; Li, Z.H.; Yan, R.X.; Li, Y.D. (2005) *Adv. Mater.*, **17**, 2119.

109. Zhang, F.; Wan, Y.; Yu, T.; Zhang, F.Q.; Shi, Y.F.; Xie, S.H.; Li, Y.G.; Xu, L.; Tu, B.; Zhao, D.Y. (2007) *Angew. Chem. Int. Ed.*, **46**, 7976.

110. Zhang, F.; Zhao, D.Y. (2009) *ACS Nano*, **3**, 159.

111. Meyssamy, H.; Riwotzki, K.; Kornowski, A.; Naused, S.; Haase, M. (1999) *Adv. Mater.*, **11**, 840.

112. Riwotzki, K.; Haase, M. (1998) *J. Phys. Chem. B*, **102**, 10129.

113. Haase, M.; Riwotzki, K.; Meyssamy, H.; Kornowski, A. (2000) *J. Alloy. Compd.*, **303–304**, 191.

114. Zhang, Y.W.; Yan, Z.G.; You, L.P.; Si, R.; Yan, C.H. (2003) *Eur. J. Inorg. Chem.*, **10**4099.

115. Yan, Z.G.; Zhang, Y.W.; You, L.P.; Si, R.; Yan, C.H. (2004) *J. Cryst. Growth*, **262**, 408.

116. Yan, R.X.; Sun, X.M.; Wang, X.; Peng, Q.; Li, Y.D. (2005) *Chem. Eur. J.*, **11**, 2183.

117. Bu, W.B.; Chen, H.R.; Hua, Z.L.; Liu, Z.C.; Huang, W.M.; Zhang, L.X.; Shi, J.L. (2004) *Appl. Phys. Lett.*, **85**, 4307.

118. Bu, W.B.; Zhang, L.X.; Hua, Z.L.; Chen, H.R.; Shi, J.L. (2007) *Cryst. Growth Des.*, **7**, 2305.

119. Guan, M.Y.; Sun, J.H.; Han, M.; Xu, Z.; Tao, F.F.; Yin, G.; Wei, X.W.; Zhu, J.M.; Jiang, X.Q. (2007)*Nanotechnology*, **18**, 1.

120. Huo, Z.Y.; Chen, C.; Chu, D.R.; Li, H.H.; Li, Y.D. (2007) *Chem. Eur. J.*, **13**, 7708.

121. Li, G.C.; Chao, K.; Peng, H.R.; Chen, K.Z. (2008) *J. Phys. Chem. C*, **112**, 6228.

122. Jia, C.J.; Sun, L.D.; Luo, F.; Jiang, X.C, Wei, L.H.; Yan, C.H. (2004) *Appl. Phys. Lett.*, **84**, 5305.

123. Jia, C.J.; Sun, L.D.; You, L.P.; Jiang, X.C, Luo, F.; Pang, Y.C.; Yan, C.H. (2005) *J. Phys. Chem. B*, **109**, 3284.

124. Chen, D.; Xu, R. (1998) *Mater. Res. Bull.*, **33**, 409.

125. Li, K.W.; Wang, Y.; Wang, H.; Zhu, M.; Yan, H. (2006) *Nanotechnology*, **17**, 4863.

126. Moon, J. (2001) *J. Am. Ceram. Soc.*, **84**, 2531.

127. Feldmann, C. (2003) *Adv. Funct. Mater.*, **13**, 101.

128. Feldmann, C. (2005) *Solid State Sci.*, **7**, 868.

129. Ho, C.M.; Yu, J.C.; Kwong, T.; Mak, A.C.; Lai, S.Y. (2005) *Chem. Mater.*, **17**, 4514.

130. Zhou, H.P.; Zhang, Y.W.; Mai, H.X.; Sun, X.; Liu, Q.; Song, W.G.; Yan, C.H. (2008) *Chem. Eur. J.*, **14**, 3380.

131. Wang, D.S.; Xie, T.; Peng, Q.; Zhang, S.Y.; Chen, J.; Li, Y.D. (2008) *Chem. Eur. J.*, **14**, 2507.

132. Ahniyaz, A.; Sakamoto, Y.; Bergstrom, L. (2008) *Cryst. Growth Des.*, **8**, 1798.

133. Gu, H.; Soucek, M.D. (2007) *Chem. Mater.*, **19**, 1103.

134. Yu, T.Y.; Joo, J.; Park, Y.I.; Hyeon, T. (2005) *Angew. Chem. Int. Ed.*, **44**, 7411.

135. Bazzi, R.; Flores-Gonzalez, M.A.; Louis C.; Lebbou, K.; Dujardin, C.; Brenier, A.; Zhang, W.; Tillement, O.; Bernstein, E.; Perriat, P. (2003) *J. Lumin.*, **102–103**, 445.

136. Bazzi, R.; Flores, M.A.; Louis, C.; Lebbou, K.; Zhang, W.; Dujardin, C.; Roux, S.; Mercier, B.; Ledoux, G.; Bernstein, E.; Perriat, P.; Tillement, O. (2004) *J. Colloi. Interf. Sci.*, **273**, 191.

137. Pinna, N.; Garnweitner, G.; Beato, P.; Niederberger, M.; Antonietti, M. (2005) *Small*, **1**, 112.

138. Karmaoui, M.; Ferreira, R.A.S.; Mane, A.T.; Carlos, L.D.; Pinna, N. (2006) *Chem. Mater.*, **18**, 4493.

139. Karmaoui, M.; Mafra, L.; Ferreira, R.A.S.; Rocha, J.; Carlos, L.D.; Pinna, N. (2007) *J. Phys. Chem. C*, **111**, 2539.

140. Pinna, N. (2007) *J. Mater. Chem.*, **17**, 2769.

141. Cao, Y.C. (2004) *J. Am. Chem. Soc.*, **126**, 7456.

142. Si, R.; Zhang, Y.W.; You, L.P.; Yan, C.H. (2005) *Angew. Chem. Int. Ed.*, **44**, 3256.

143. Si, R.; Zhang, Y.W.; Zhou, H.P.; Sun, L.D.; Yan, C.H. (2007) *Chem. Mater.*, **19**, 18.

144. Yu, T.K.; Joo, J.; Park, Y.I.; Hyeon, T. (2006) *J. Am. Chem. Soc.*, **128**, 1786.

145. Wang, H.Z.; Uehara, M.; Nakamura, H.; Miyazaki, M.; Maeda, H. (2005) *Adv. Mater.*, **17**, 2506.

146. Zhang, Y.X.; Guo, J.; White, T.; Thatt, T.; Tan, Y.; Xu, R. (2007) *J. Phys. Chem. C*, **111**, 7893.

147. Han, M.; Shi, N.E.; Zhang, W.L.; Li, B.J.; Sun, J.H.; Chen, K.J.; Zhu, J.M.; Wang, X.; Xu, Z. (2008) *Chem. Eur. J.*, **14**, 1615.

148. Paek, J.; Lee, C.H.; Choi, J.; Choi, S.Y.; Kim, A.; Lee, J.W.; Lee, K. (2007) *Cryst. Growth Des.*, **7**, 1378.

149. Wu, L.M.; Seo, D.K. (2004) *J. Am. Chem. Soc.*, **126**, 4676.

150. Tanaka, K.; Tatehata, N.; Fujita, K.; Hirao, K. (2001) *J. Appl. Phys.*, **89**, 2213.

151. Mirkovic, T.; Hines, M,A.; Nair, P.S.; Scholes, G.D. (2005) *Chem. Mater.*, **17**, 3451.

152. Huxter, V,M.; Mirkovic, T.; Nair, P.S.; Scholes, G.D. (2008) *Adv. Mater.*, **20**, 2439.

153. Zhao, F.; Sun, H.L.; Gao, S.; Su, G. (2005) *J. Mater. Chem.*, **15**, 4209.

154. Zhao, F.; Sun, H.L.; Su, G.; Gao, S. (2006) *Small*, **2**, 244.

155. Regulacio, M.D.; Bussmann, K.; Lewis, B.; Stoll, S.L. (2006) *J. Am. Chem. Soc.*, **128**, 11173.

156. Regulacio, M.D.; Kar, S.; Zuniga, E.; Wang, G.B.; Dollahon, N.R.; Yee, G.T.; Stoll, S.L. (2008) *Chem. Mater.*, **20**, 3368.

157. Hasegawa, Y.; Afzaal, M.; Brien, P.O.; Wada, Y.; Yanagida, S. (2005) *Chem. Commun.*, 242.

158. Yu, S.H.; Han, Z.H.; Yang, J.; Zhao, H.Q.; Yang, R.Y.; Xie, Y.; Qian, Y.T.; Zhang, Y.H. (1999) *Chem. Mater.*, **11**, 192.

159. Jiang, Y.; Wu, Y.; Xie, Y.; Qian, Y.T. (2000) *J. Am. Ceram. Soc.*, **83**, 2628.

160. Wang, X.; Sun, X.M.; Yu, D.P.; Zou, B.S.; Li, Y.D. (2003) *Adv. Mater.*, **15**, 1442.

161. Wang, X.; Li, Y.D. (2003) *Chem. Eur. J.*, **9**, 5627.

162. Zhao, F.; Gao, S. (2008) *J. Mater. Chem.*, **18**, 949.

163. Zhao, F.; Yuan, M.; Zhang, W.; Gao, S. (2006) *J. Am. Chem. Soc.*, **128**, 11758.

164. Wang, Z.L.; Quan, Z.W.; Jia, P.Y.; Lin, C.K.; Luo, Y.; Chen, Y.; Fang, J.; Zhou, W.; O'Connor, C. J.; Lin, J. (2006) *Chem. Mater.*, **18**, 2030.

165. Qin, R.F.; Song, H.W.; Pan, G.H.; Bai, X.; Dong, B.; Xie, S.H.; Liu, L.N.; Dai, Q.L.; Qu, X.S.; Ren, X.G.; Zhan, H.F. (2009) *Cryst. Growth Des.*, **9**, 1750.

166. Zhang, Y.W.; Sun, X.; Si, R.; You, L.P.; Yan, C.H. (2005) *J. Am. Chem. Soc.*, **127**, 3260.

167. Sun, X.; Zhang, Y.W.; Du, Y.P.; Yan, Z.G.; Si, R.; You, L.P.; Yan, C.H. (2007) *Chem. Eur. J.*, **13**, 2320.

168. Du, Y.P.; Zhang, Y.W.; Sun, L.D.; Yan, C.H. (2008) *J. Phys. Chem. C*, **112**, 405.

169. Du, Y.P.; Zhang, Y.W.; Sun, L.D.; Yan, C.H. (2009) *J. Am. Chem. Soc.*, **131**, 3162.

170. Zhuravleva, N.G.; Eliseev, A.A.; Sapoletova, N.A.; Lukashin, A.V.; Kynast, U.; Tretyakov Yu, D. (2005) *Mater. Sci. Eng. C*, **25**, 549.

171. Mai, H.X.; Zhang, Y.W.; Si, R.; Yan, Z.G.; Sun, L.D.; You, L.P.; Yan, C.H. (2006) *J. Am. Chem. Soc.*, **128**, 6426.

172. Du, Y.P.; Zhang, Y.W.; Sun, L.D.; Yan, C.H. (2009) *Dalton Trans.*, 8574.

173. Riwotzki, K.; Meyssamy, H.; Kornowski, A.; Haase, M. (2000) *J. Phys. Chem. B*, **104**, 2824.

174. Riwotzki, K.; Meyssamy, H.; Schnablegger, H.; Kornowski, A.; Haase, M. (2001) *Angew. Chem. Int. Ed.*, **40**, 573.

175. Lehmann, O.; Kompe, K.; Haase, M. (2004) *J. Am. Chem. Soc.*, **126**, 14935.

176. Heer, S.; Lehmann, O.; Haase, M.; Gudel, H.U. (2003) *Angew. Chem. Int. Ed.*, **42**, 3179.

177. Kompe, K.; Borchert, H.; Storz, J.; Lobo, A.; Adam, S.; Moller, T.; Haase, M. (2003) *Angew. Chem. Int. Ed.*, **42**, 5513.

178. Mai, H.X.; Zhang, Y.W.; Sun, L.D.; Yan, C.H. (2007) *Chem. Mater.*, **19**, 4514.

179. Liao, X.H.; Zhu, J.M.; Zhu, J.J.; Xu, J.Z.; Chen, H.Y. (2001) *Chem. Commun.*, 937.

180. Godinho, M.; Ribeiro, C.; Longo, E.; Leite, E.R. (2008) *Cryst. Growth Des.*, **8**, 384.

181. Panda, A.B.; Glaspell, G.; El-Shall, M.S. (2007) *J. Phys. Chem. C*, **111**, 1861.

182. Hasegawa, Y.; Okada, Y.; Kataoka, T.; Sakata, T.; Mori, H.; Wada, Y. (2006) *J. Phys. Chem. B*, **110**, 9008.

183. Ma, L.; Chen, W.X.; Zheng, Y.F.; Zhao, J.; Xu, Z.D. (2007) *Mater. Lett.*, **61**, 2765.

184. Masui, T.; Fujiwara, K.; Machida, K.; Adachi, G.; Sakata, T.; Mori, H. (1997) *Chem. Mater.*, **9**, 2197.

185. Bumajdad, A.; Zaki, M.I.; Eastoe, J.; Pasupulety, L. (2004) *Langmuir*, **20**, 11223.

186. Sathyamurthy, S.; Leonard, K.J.; Dabestani, R.T.; Paranthaman, M.P. (2005) *Nanotechnology*, **16**, 1960.

187. Lian, H.; Zhang, M.; Liu, J.; Ye, Z.; Yan, J.; Shi, C.S. (2004) *Chem. Phys. Lett.*, **395**, 362.

188. Lemyre, J.L.; Ritcey, A.M. (2005) *Chem. Mater.*, **17**, 3040.

189. Wang, G.F.; Qin, W.P.; Zhang, J.S.; Cao, C.Y.; Wang, L.L.; Wei, G.D.; Zhu, P.F.; Kim, R. (2008) *J. Phys. Chem. C*, **112**, 12161.

190. Cao, M.H.; Hu, C.W.; Wu, Q.Y.; Guo, C.X.; Qi, Y.J.; Wang, E.B. (2005) *Nanotechnology*, **16**, 282.

191. Xing, Y.; Li, M.; Davis, S.A.; Mann, S. (2006) *J. Phys. Chem. B*, **110**, 1111.

192. Ghosh, P.; Oliva, J.; De la Rosa, E.; Haldar, K.K.; Solis, D.; Patra, A. (2008) *J. Phys. Chem. C*, **112**, 9650.

193. Zhang, D.S.; Fu, H.X.; Shi, L.Y.; Pan, C.S.; Li, Q.; Chu, Y.L.; Yu, W.J. (2007) *Inorg. Chem.*, **46**, 2446.

194. Zhu, L.; Li, J.Y.; Li, Q.; Liu, X.D.; Meng, J.; Cao, X.Q. (2007) *Nanotechnology*, **18**, 055604.

195. Zhu, L.; Li, Q.; Liu, X.D.; Li, J.Y.; Zhang, Y.F.; Meng, J.; Cao, X.Q. (2007) *J. Phys. Chem. C*, **111**, 5898.

196. Yu, C.C.; Yu, M.; Li, C.X.; Liu, X.M.; Yang, J.; Yang, P.P.; Lin, J. (2008) *J. Solid State Chem.*, **182**, 339.

197. Li, Z.X.; Li, L.L.; Yuan, Q.; Feng, W.; Xu, J.; Sun, L.D.; Song, W.G.; Yan, C.H. (2008) *J. Phys. Chem. C*, **112**, 18405.

198. Chen, C.; Sun, L.D.; Li, Z.X.; Li, L.L.; Zhang, J.; Zhang, Y.W.; Yan, C.H. (2010) *Langmuir,*, **26**, 8797.

199. Zhang, C.; Chen, J.; Zhou, Y.C.; Li, D.Q. (2008) *J. Phys. Chem. C*, **112**, 10083.

200. Bühler, G.; Feldmann, C. (2006) *Angew. Chem. Int. Ed.*, **45**, 4864.

201. Xia, B.; Lenggoro, I.W.; Okuyama, K. (2001) *J. Mater. Chem.*, **11**, 2925.

202. Hu, C.G.; Zhang, Z.W.; Liu, H.; Gao, P.X.; Wang Z.L. (2006) *Nanotechnology*, **17**, 5983.

203. Huang, H.; Gur, T.M.; Saito, Y.; Prinz, F. (2006) *Appl. Phys. Lett.*, **89**, 143107.

204. Sun, X.M.; Liu, J.F.; Li Y.D. (2006) *Chem. Eur. J.*, **12**, 2039.

205. Karakoti, A.S.; Kuchibhatla, S.; Baer, D.R.; Thevuthasan, S.; Sayle, D.C.; Seal, S. (2008) *Small*, **4**, 1210.

206. Thongchant, S.; Hasegawa, Y.; Wada, Y.; Yanagida, S. (2003) *J. Phys. Chem. B*, **107**, 2193.

207. Kataoka, T.; Tsukahara, Y.; Hasegawa, Y.; Wada, Y. (2005) *Chem. Commun.*, 6038.

Chapter 2

Structural Control and Surface Modifications of Rare Earth Nanomaterials

Gautom Kumar Das and Timothy T.Y. Tan
School of Chemical and Biomedical Engineering, Nanyang Technological University, 637459, Singapore
tytan@ntu.edu.sg

2.1 Introduction

Controlled synthesis of functional nanomaterials with tunable composition, phase, morphology, and surface properties and their subsequent utilization as building blocks for the fabrication of nanodevices have been attracting significant research interests in material science and technology [1–6]. Nanoscale materials show unique properties compared to that of bulk materials. For example, gold at nanoscale (1.6 nm) shows much lower melting point (350°C) than in bulk (1064°C) [2]. In principle, the electron confinement by a nanocrystal provides the most powerful means to manipulate the electronic, optical, and magnetic properties of a solid material [1]. This explains why nanocrystals have been the preliminary source of discovery of quantum size effects such as superparamagnetism [7, 8], quantized excitation [9, 10], metal–insulator transition [11],

Rare Earth Nanotechnology
Edited by Timothy T.Y. Tan

ISBN 978-981-4316-30-9 (Hardcover), 978-981-4364-20-1 (eBook)
www.panstanford.com

and Coulomb blockade [12]. Thus, nanomaterials are at the unique position to bridge the understanding between atoms and their bulk counterparts.

The emergence of rare earth nanomaterials research has arisen from the potential of these materials in a wide range of applications such as biomedical, catalysis, and optoelectronics [13–22]. In biomedical field, trivalent rare earth ions doped nanomaterials are extremely promising as biolabels due to their non-toxicity [23], paramagnetism [14, 16], and ability to produce upconversion emission [24–28]. The upconverting rare earth nanomaterials produce emissions in the range of visible to near-infrared (NIR) wavelengths [21] of which the latter is an ideal window for deep tissue optical imaging [29, 30]. Unlike semiconductor quantum dots, rare earth doped nanomaterials exhibit size-independent luminescence, which originates from intra-configurational $4f^n$ electron transitions within the localized doped rare earth ions.

In view of rapidly growing number of reports in this domain, it is necessary to address the aspect of controlled synthesis of rare earth nanomaterials and the associated structural and functional modifications. Synthesis of inorganic nanocrystals with controlled morphology with desired phase and surface properties have been reviewed comprehensively elsewhere for most of the metal nanomaterials [1–5, 31, 32]. Careful control of nucleation and growth processes is required to achieve crystallographic control and fabrication of a low-dimensional structure. Generally, solid-state and solution-based routes have been employed to synthesize a wide range of rare earth based nanomaterials, including oxides, oxysulfides, vanadites, phosphates, and fluorides [26, 27, 33–35]. Solution-based routes have the versatility in terms of synthesis and post-synthesis modifications and ease of control in a wide range of temperatures. Most commonly, metal salt precursors are reduced in solution (aqueous or otherwise) in the presence of a stabilizing agent to prevent aggregation or improve the chemical stability of the nanomaterials. The reactions of the nanomaterial formations are governed by thermodynamic (e.g., temperature, reduction potential), kinetic (e.g., concentration of the reactants, solubility, diffusion, reaction rate) as well as intrinsic (e.g., ionic radii, polarizability) properties that are intimately and intricately associated [36–40]. On the other hand, surface modification is a criticial step for high-value applications involving nanoparticles.

In this chapter, we will limit our discussions to controlled synthesis of rare earth nanostructures only. The concept of nucleation and crystallization will be illustrated first, followed by a review of dimensionally and phase-controlled rare earth nanostructures and methods reported for their surface modification.

2.2 Nucleation, Seed, and Crystals

Nucleation is the very first stage of a crystallization process. In a typical synthesis process, precursor compounds are decomposed or reduced to generate zero-valent atoms (i.e., the building blocks of a metal nanocrystal). In spite of significant technological advancement, we have only achieved limited understanding and, therefore, control of the nucleation process [41]. One of the main constrains is the lack of experimental tools capable of capturing and monitoring nucleation. Nuclei are formed by a cluster of few atoms and/or ions, and by the time a crystal is visible to electron microscopes, it has already grown beyond the nucleation stage.

Two main nucleation routes have been proposed, namely, decomposition and reduction routes, by which nucleation is believed to occur. Based on a study of the solution-phase synthesis of sulfur colloids, LaMer *et al.* [42] proposed a mechanism in which nucleation is expected to occur for the decomposition route. Figure 2.1 schematically illustrates the mechanism of crystal growth. It was proposed that the concentration of metal atoms steadily increases with time as the precursor is decomposed. As the concentration of the atoms reaches a point of supersaturation, the atoms start to aggregate into small clusters (i.e., nuclei) via homogenous (self-) nucleation and continue to grow until the concentration of metal atoms in solution drops. No additional nuclei will be formed when the concentration of atoms drops below the level of minimum supersaturation. However, with a continuous supply of atoms via constant precursor decomposition, the nuclei will grow into nanocrystals of increasingly larger size until an equilibrium state is reached between the atoms on the surface of the nanocrystals and the atoms in the solution. Aggregation of nuclei results in bigger crystals as well [43].

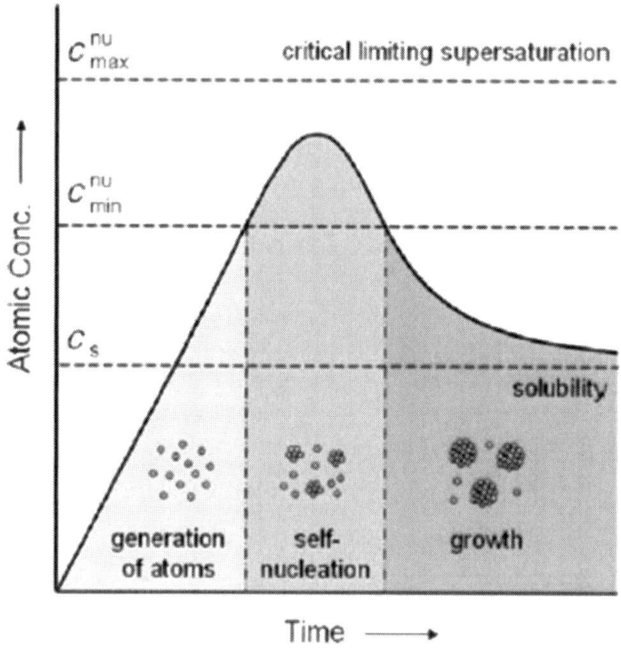

Figure 2.1 A crystal-growth diagram illustrates the generation of atoms, nucleation, and subsequent growth [1].

For the reduction route, the precursor compound is in a higher oxidation state than the atomic species. Either the precursor compounds are reduced to zero-valent atoms and agglomerate further into nanocrystals, or the unreduced metal species form nuclei prior to reduction [1]. Structural fluctuations become energetically unfavorable beyond a critical size of the atom cluster and then it locks into a well-defined structure, giving rise to a seed crystal [44]. In general, the seeds may be in the form of a single-crystal, singly twinned, or multiply twinned structure as shown in Fig. 2.2. In a typical synthesis, one or more type of seeds may coexist. The key to obtaining the desired type of nanocrystals and exclusion of others is generally dependent on precise control of the synthesis parameters, especially those of thermodynamics and kinetics. Figure 2.2 shows general reaction pathways that lead to seeds formation, and finally to fully formed nanostructures of various morphologies. These pathways are illustrative for homogeneous nucleation. However, for heterogeneous nucleation, shape control is less stringent. Core–shell nanostructures are examples of products of heterogeneous nucleation. The activation energy for metal reduction onto a preformed particle is significantly lower than that for homogenous nucleation of seed particles in a solution [1]. Heterogeneous nucleation process can be considered an overgrowth process where seed particles are added to a growth medium to facilitate the reduction of metal ions.

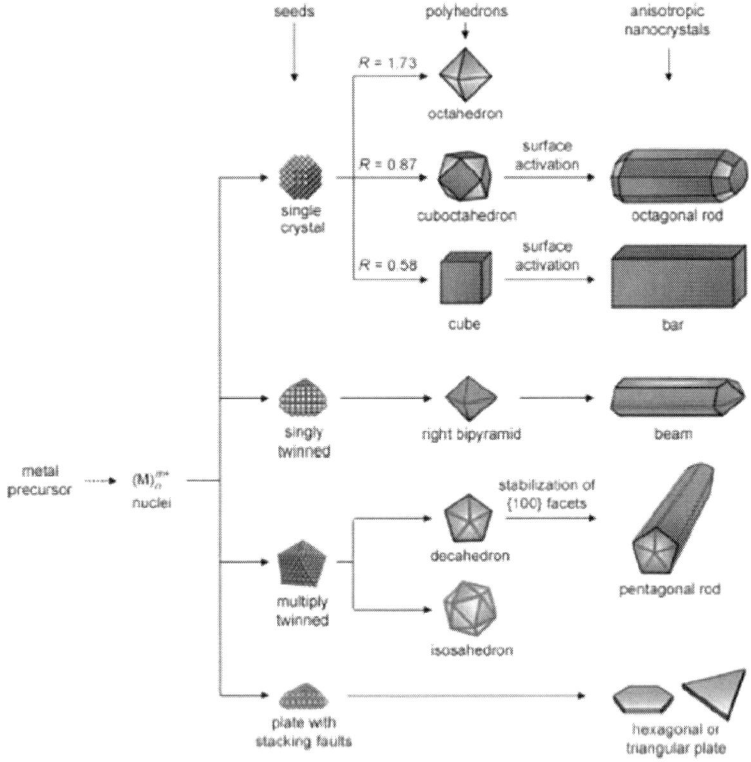

Figure 2.2 Reaction pathways that lead to seeds to different shaped nanostructures [1].

2.3 Dimensionally Controlled Rare Earth Nanomaterials

Controlled synthesis of rare earth nanomaterials will be presented in this section. The reported rare earth nanomaterials have been classified by their dimensional features such as zero-, one-, and two-dimensional (D) structures. Reported core–shell, hollow, and complex structures and their synthesis will also be presented.

Basic geometrical motifs of reported rare earth nanomaterials are displayed in Fig. 2.3. The nanostructures that have equal or similar dimensions in all directions are defined as 0D nanostructures (e.g., nanospheres), while 1D nanostructures have one elongated direction (e.g., nanowire/nanotubes), and 2D nanostructures have one flatten structure (e.g., nanoplates).

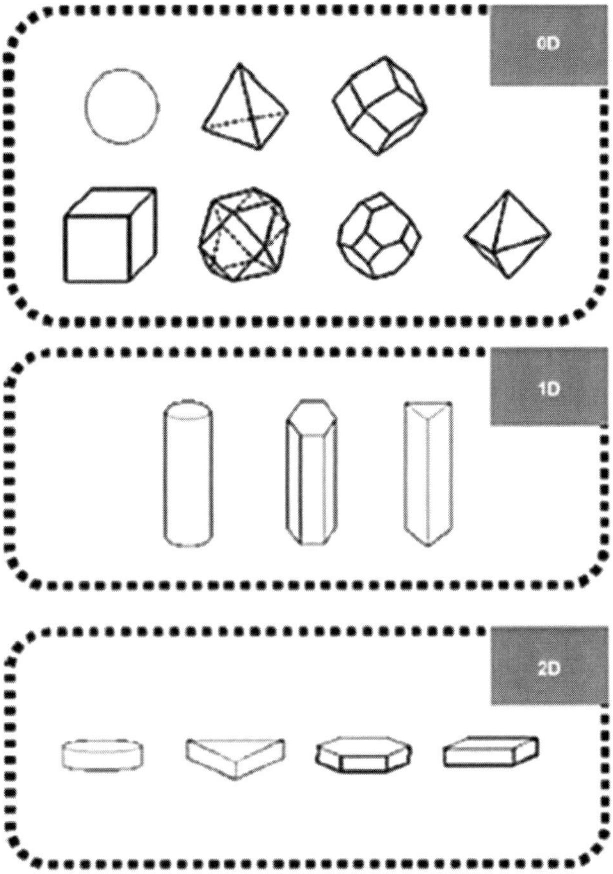

Figure 2.3 Basic geometric motifs of the inorganic nanostructures: 0D spheres, cubes, and polyhedrons; 1D rods and wires; 2D discs, prisms, and plates [2].

Generally, dimensionally controlled nanostructures have attracted significant interest for a wide range of materials, some of which also demonstrated fascinating properties. For example, Ag and Au nanocrystals of different shapes possess unique optical scattering responses [45–47]. Although highly symmetric spherical particles exhibit a single scattering peak, anisotropic shapes — such as rods [47], triangular prism [45], and cubes [46] — of these materials exhibit multiple scattering peaks in the visible wavelengths due to highly localized charge polarizations at corners and edges. This suggests that control of nanocrystals shape could offer a strategy for optical tuning. Yet, chemical reactivity is highly dependent on surface morphology, the bonding facets of the nanocrystal, the number of the step edges and kink sites, as well as the surface-to-volume ratio. By exploiting these properties, El-Sayed *et al.* [48] showed that shape-controlled Pt and Pd nanocrystals can be used to achieve highly selective catalysis. In another study, Huynh *et al.* [49] showed that by integrating 1D semiconducting CdSe nanorods with P3HT hole transporting polymers, photovoltaic devices (i.e., solar cells) can be fabricated. In another work, the same group demonstrated the development of a new type of single-electron transistor (SET) using CdSe tetrapods [50]. Therefore, shape-controlled nanomaterials do not have merely aesthetic appeal but offer huge promise in catalysis, electronic, as well as biomedical industries. In the following sections, however, we will limit our discussion to representative dimensionally controlled rare earth nanostructures, their synthesis, and mechanism of formations.

2.3.1 Zero-Dimensional (0D) Rare Earth Nanostructures

Among the 0D nanocrystals, spheres are the most basic and symmetric motif. Cubes, truncated cubes, truncated octahedral, and octahedral structures are other typical shapes. There have been a great number of reports on various 0D rare earth nanoparticles. However, only a few representative works will be highlighted in this section.

Rare earth fluorides (REF_4), an extensively studied class of rare earth materials, will be used to illustrate different forms of 0D nanostructures. Li *et al.* reported a series of REF_4 nanostructures using different methods and techniques [51, 52]. The group developed a hydrothermal synthetic strategy based on liquid-solid-solution growth, which was employed to synthesize a range of REF_4 nanomaterials. Figure 2.4 shows the TEM images of the hexagonal- or orthorhombic-phase polyhedron of fluorides. Alkali-lioleate, linoleate acid, and ethanol were used with rare earth nitrate precursors at a reaction temperature of 100–200°C. The resulting nanocrystals were varied in shape and size but generally monodisperse and crystalline. The authors also investigated the dependence of nanocrystal shapes on the rare earth ionic radius and proposed that the growth of nanocrystals changed with ionic size.

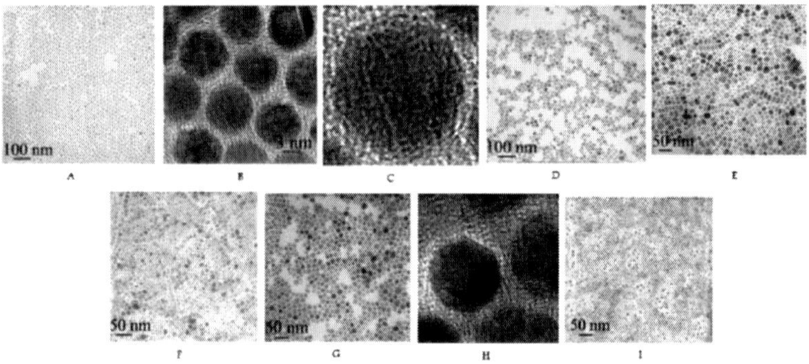

Figure 2.4 TEM images of (A, B, C) NaYF4, (D) CeF3, (E) PrF4, (F) NdF3, (G,H) LaF3, (I) NaYb2F7 [52].

Using a microemulsion method, Lemyre and Ritcey [53] demonstrated the synthesis of a range of 0D fluorides nanomaterials. YCl_3 and NH_4HF_2 were used as precursors and dissolved in the microemulsion of water, Igepal CO520, and cyclohexane. The resultant 0D nanoparticles showed high crystallinity, monodispersity, and different morphologies (Fig. 2.5). Surfactant-to-water ratio and precursor concentrations played a critical role in shape and size control of these materials. Similar 0D spherical Y_2O_3 has been reported by Arriagada *et al.* [54] using a microemulsion method.

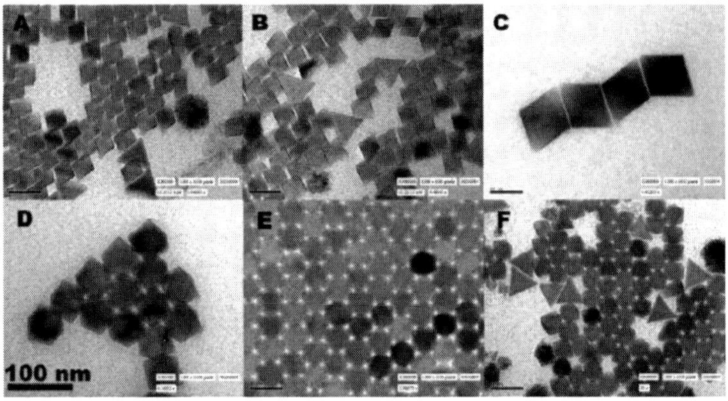

Figure 2.5 Different morphologies of REF_3 nanoparticles obtained using a microemulsion method: (A), (B) mixture of hexagonal, triangular, and quadrilateral YF_3 nanoparticles, (C) quadrilateral, (D) hexagonal, (E) self-assembly of hexagonal nanoparticles, and (F) ErF_3 particles [53].

Among others, Ceria (CeO_2)-based nanomaterials are widely investigated and reported in the form of almost all the morphologies. Ceria has a wide range of applications, including conversion catalysts, solar cells, fuel cells, gates for metal oxide semiconductor devices, and phosphors [55–59]. However, only a few of the reported ceria 0D nanoparticles will be discussed here. Yang *et al.* [60] reported a thermal decomposition method to produce ordered and monodisperse nanocubes (Fig. 2.6a). By changing the concentration of the reactants, the amount of stabilizing agents (oleic acid) and the water/toluene ratio in the reaction system, the size and morphology could be tuned. Using a hydrothermal method and $Ce(NO_3)_3$, Yan *et al.* [61] reported nano-octahedrons of CeO_2 (Fig. 2.6b). Careful control of experimental conditions such as time, temperature, additives, pH value, and concentration of the precursors provided a way to grow CeO_2 nanorods and nano-octahedrons. There are a good number of reports that demonstrated controlled synthesis of 0D rare earth nanomaterials with many exciting properties such as YPO_4 (Fig. 2.6c) as multicolor upconversion emitter [21], ultrasmall Gd_2O_3 (Fig. 2.6d) as multimodal imaging nanoprobes [25].

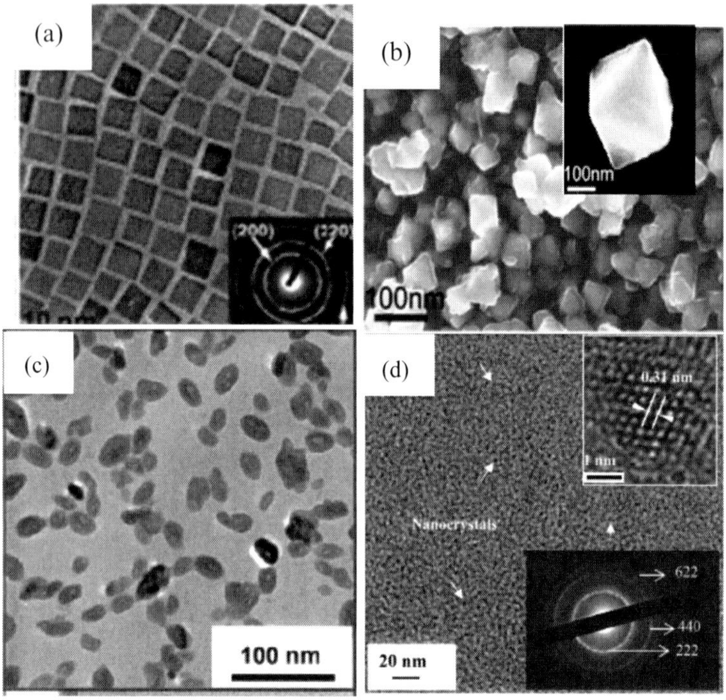

Figure 2.6 0D rare earth nanoparticles of different morphology: (a) nanocubes [60], (b) nano-octahedrons of CeO_2 [61], (c) nearly spherical YVO_4 [21], and (d) ultrasmall spherical Gd_2O_3 nanocrystals [25].

2.3.2 One-Dimensional (1D) Rare Earth Nanostructures

In the past decade, a variety of synthetic strategies have been examined and demonstrated as a "bottom-up" approach for fabricating 1D nanostructures [62]. These strategies can be generally classified as (Fig. 2.7) [32]: (a) the use of intrinsically anisotropic crystallographic structure of a nanomaterial to accomplish 1D growth; (b) the introduction of a liquid–solid interface to reduce the symmetry of a seed; (c) the use of various templates with 1D morphologies to direct the formation of 1D nanostructures; (d) the use of appropriate capping reagents to kinetically control the growth rates of various facets of a seed; (e) self-assembly of 0D nanostructures; and (f) a "top-down" strategy such as the size reduction of 1D microstructures.

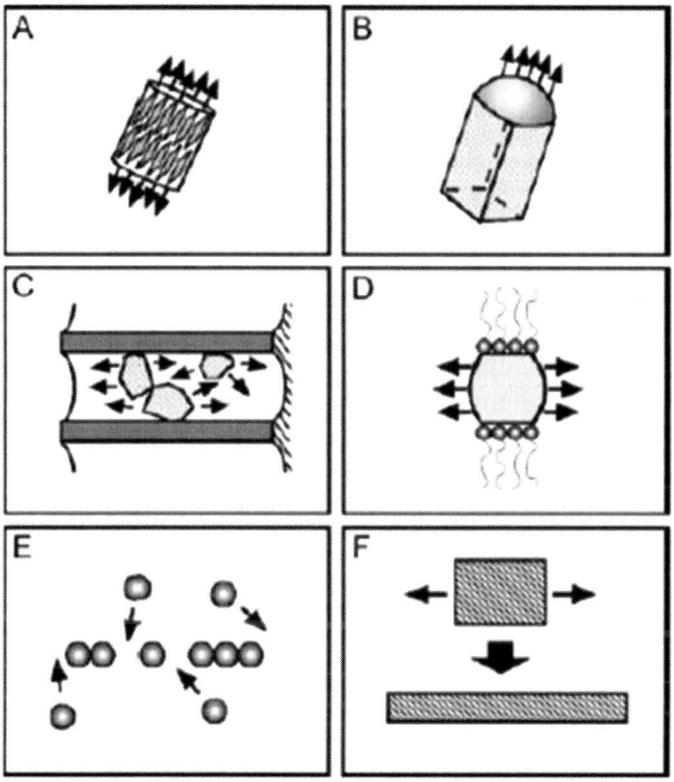

Figure 2.7 Illustration of six different strategies that have been demonstrated for the fabrication of 1D growth: (a) dictation by the anisotropic crystallographic structure of a solid; (b) confinement by liquid droplet in a vapor–liquid–solid process; (c) direction through the use of a template; (d) kinetic control provided by a capping reagent; (e) self-assembly of 0D nanostructures; and (f) size reduction of a 1D microstructure [32].

Nanorods, nanowires, and nanobelts are the typical examples of the 1D rare earth nanostructures. The rare earth nanostructures show strong dependence on the phase and crystal structure, while ion mobility and solubility of the rare earth ions also play key role [63]. In this section, we will discuss the factors that affect the morphology of different types of materials and the synthetic parameters that affect the 1D growth of rare earth nanostructures in each system.

Wang *et al.* [64] demonstrated a hydrothermal route to obtain a range of rare earth hydroxides (RE = La, Pr, Nd, Sm, Eu, Gd, Dy,

Tb, Ho, Tm) nanowires. In a typical synthesis, rare earth hydroxides gels are precipitated in basic solutions (KOH) at room temperature and treated hydrothermally at 180°C to produce nanowires. The synthesized nanowires are shown in Fig. 2.8. The pH of the solution played a pivotal role in shape tuning. For example, at lower pH (i.e., 6–7), $Sm(OH)_3$ nanosheets of low aspect ratio were obtained (Fig. 2.8c), while a higher pH (i.e., 9–10) resulted in nanowires of high aspect ratio (Fig. 2.8d). Limited mobility of Sm^{3+} ions in higher pH was inferred to be responsible for the shape formation. The simulated crystal structure of $RE(OH)_3$ (Fig. 2.8b) provides a better insight. Along the *c*-axis, there are 1D chains consisting of alternating OH^- anions and RE^{3+} cations connected to each other. With changing pH, OH^- ligands can capture or release protons to form O^{2-} or H_2O, changing the sign and density of charges on the growing crystal faces, which ultimately change the crystal morphology. The synthesized nanowires are proposed to be used as potential biolabels.

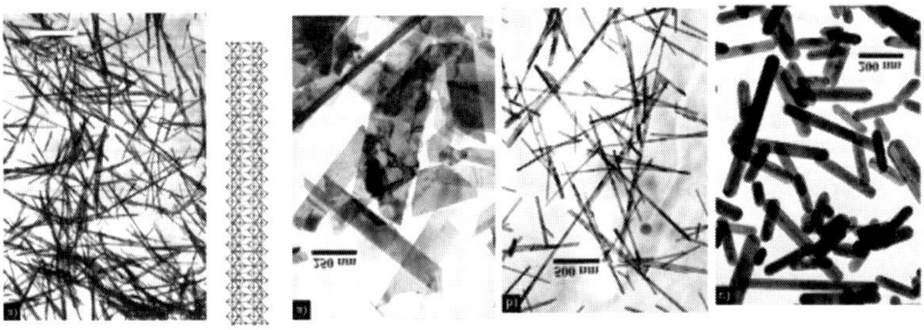

Figure 2.8 (a) TEM image of $La(OH)_3$ nanowires as a representative RE-hydroxide; inset: HRTEM image of a single $La(OH)_3$ nanowire; (b) simulated crystal mode, TEM images of $Sm(OH)_3$; (c) nanosheets (pH 6–7); (d) nanowires (pH 9–10); (e) nanorods (high alkali volume) [64].

Rare earth orthophosphates exhibit high thermal stability, low water solubility, and high refractive index and, therefore, are promising in ceramics applications. Kinetically controlled (i.e., temperature-dependent) 1D wire-like $REPO_4$ (RE = La, Ce, Pr, Nd, Sm, Eu, Gd, Tb, Dy) were reported by Fang *et al.* [65] (Fig. 2.9). At room temperature, gel-like precipitates comprising ball-like nanoparticle assemblies were formed, while subsequent treatment at 140–240°C resulted in particles, rods, and wires depending on the temperature and pH of the solution. At 140°C, pure hexagonal nanocrystals were formed, while at elevated temperature (at 240°C), the nanocrystals were converted to monoclinic nanorods and nanowires.

Figure 2.9 SEM images of the synthesized $REPO_4$ nanowires by hydrothermal treatment: (a) $PrPO_4$, (b) $SmPO_4$, (c) $GdPO_4$, and (d) $TbPO_4$. TEM images of (e) hexagonal $LaPO_4$, (f) $CePO_4$ nanowires or nanorods [65].

The use of capping ligands to control the growth rates of various facets of a seed, which leads to 1D rare earth orthovanadate ($REVO_4$) nanostructures, has been demonstrated by Luo *et al.* [66]. $REVO_4$ has potential applications in photonics and catalysis. The authors demonstrated a chelating-ligand EDTA mediated synthesis of $CeVO_4$ nanostructures using a wide range of pH (i.e., 1–14) and temperature from 140–240°C. The tuning of EDTA/Ce^{3+} ratio, pH, and temperature produced rod-like, flower-like, woolen-shaped, dumbbell-like, and even hollow spheres (Fig. 2.10). Methods reported for the controlled synthesis of 1D $REVO_4$ nanostructures include hydrothermal, reverse microemulsion, and precipitation methods [67–70]. The pH of the solution plays a critical role in the synthesis of rare earth orthovanadates as VO_4^{3-} is reversibly condensed into $V_3O_9^{3-}$ and other complex anions or precipitated as V_2O_5 in acidic solutions. Therefore, most of the aqueous syntheses have been carried out in basic conditions [71].

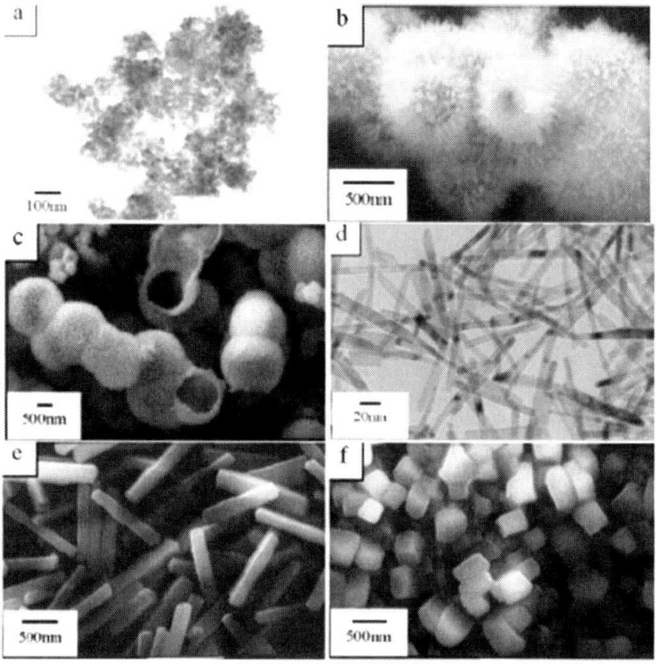

Figure 2.10 SEM and TEM images of $CeVO_4$ samples with typical shapes obtained at (a) pH 10, 180°C, 24 h, EDTA/Ce = 0; (b) pH 1, 180°C, 24 h, EDTA/Ce = 1.5; (c) pH 3, 180°C, 24 h, EDTA/Ce = 1.5; (d) pH 10, 180°C, 24 h, EDTA/Ce = 1.0; (e) pH 10, 180°C, 24 h, EDTA/Ce = 3.0; and (f) pH 10, 220°C, 24 h, EDTA/Ce = 3.0 [66].

Self-assembly of rare earth nanostructures has been demonstrated by Das *et al.* [25, 38, 39]. The group developed a method to synthesize Y_2O_3 and Gd_2O_3 nanorods for optical imaging and optical magnetic resonance imaging, respectively. The oleic acid mediated thermal decomposition of rare earth precursors resulted in uniform, monodisperse nanorods. The synthesis conditions such as temperature, precursor ratio, time, and type of the liganding solvents were tuned to produce 1D nanorods. Representative TEM images of Gd_2O_3 nanorods are shown in Fig. 2.11. The authors proposed a self-assembly mechanism of the nanocrystals through oriented attachments, which resulted in 1D nanorods (Fig. 2.12). In the oriented attachment, particles appeared to fuse end-to-end along the longitudinal axis and form linear chains. The attachment leads to a lowering of the surface energy after the elimination of highly curved surfaces of individual spheroids and is an enthalpy favorable process. The 1D nanostructures showed strong luminescence and good contrast in T_1-weighted magnetic resonance imaging.

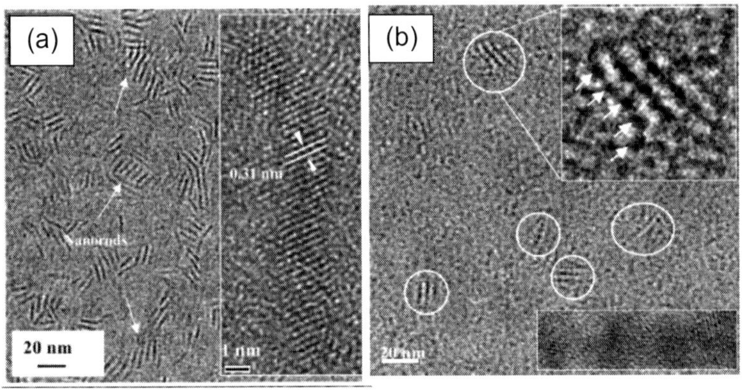

Figure 2.11 (a) TEM image of monodisperse Gd_2O_3 nanorods; inset: HRTEM image of a single nanorod. (b) TEM images of samples at early stage of growth showing the formation of pearl necklace-like chains at the time of self-assembly [25].

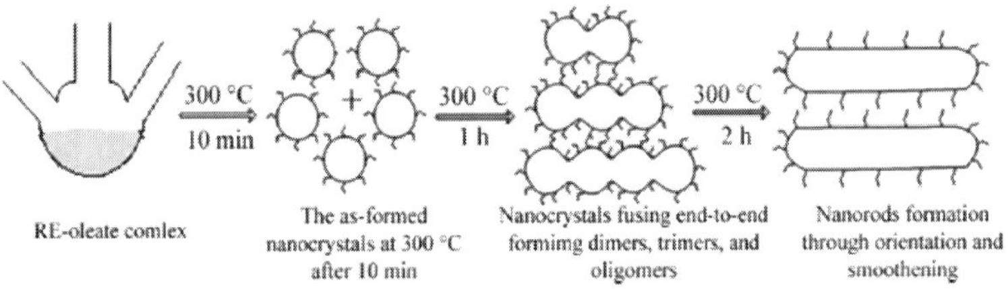

Figure 2.12 Proposed mechanism of nanorods formation by oriented attachment [25].

Rare earth fluorides, the most efficient rare earth host nanomaterials for upconversion luminescent nanomaterials, have been prepared by hydrothermal, thermal decomposition, and precipitation methods, resulting in different type of structures [27, 72–74]. For example, Mai *et al.* [74] synthesized luminescent $NaYF_4$ 1D nanorods, 2D hexagonal nanoplates, and 0D nanoparticles (Fig. 2.13). In the synthesis, oleic acid and alcohol were co-added with water and NaOH. The authors also investigated the mechanism of shape formations. They proposed that alcohol chain length, reaction time, temperature, and the ratio of F^- to Y^{3+} were determinant in producing different shape and size.

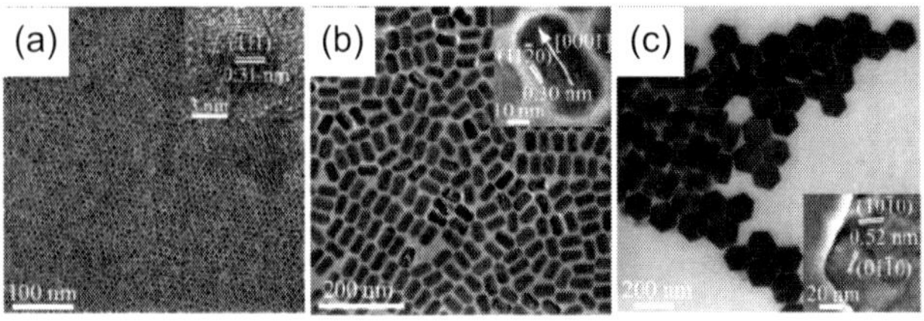

Figure 2.13 TEM and HRTEM (inset) images of: (a) α-$NaEuF_4$ hexagonal nanocrystals, (b) β-$NaYF_4$ nanorods, and (c) β-$NaYF_4$ nanoplates [75].

The above examples are by no means exhaustive but representative among the vast amount of literatures on rare earth 1D nanostructures.

2.3.3 Two-Dimensional (2D) Rare Earth Nanostructures

In a kinetically driven growth regime, the growth of 1D nanorod is typically promoted at a faster rate along a specific direction (e.g., *z*-axis; Fig. 2.14a). On the other hand, when growth along an axis is inhibited and preferential growth occurs along the other two axes (e.g., *x*- and *y*-axes; Fig. 2.14b), the formation of disc-shaped nanocrystals is facilitated. Unlike 1D rods, the formation of 2D disc-shaped nanocrystals is not very common in colloidal systems. Puntes *et al.* [76] is one of the first groups to report a strategy for the synthesis of 2D nanostructures of metallic cobalt nanodiscs.

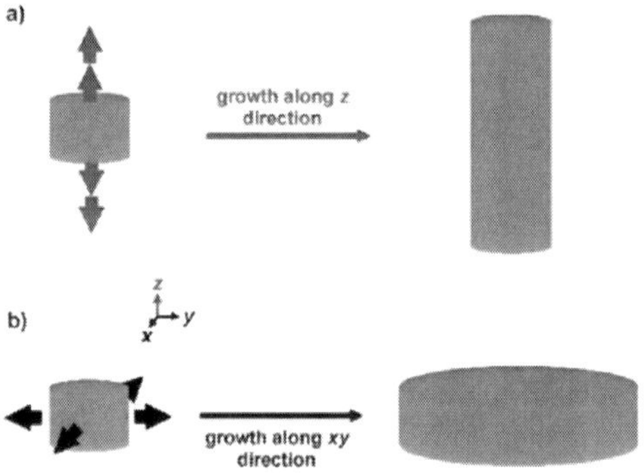

Figure 2.14 1D rod versus 2D disc growth. (a) Preferential growth along one direction (e.g., *z*-axis) of a seed results in rod-shaped nanocrystals; (b) growth along two directions (*x*- and *y*-planes) leads to the formation of disc-shaped nanocrystals [76].

Two-dimension rare earth nanostructures are usually obtained using high boiling point solvents through precipitation route or in the presence of strong chelating reagents such as oleic acid through hydrothermal routes [74]. Hydrothermal syntheses without surfactants have also been demonstrated for obtaining 2D nanostructures [77]. The synthesis of 2D nanostructures needs very selective and tight control of synthesis parameters. Some representative rare earth 2D nanostructures are discussed in this section.

Cao [78] reported the synthesis of Gd_2O_3 2D nanoplates using a thermal decomposition route in a mixture of oleic acid, oleylamine, and octadecylamine. Based on electron diffraction (ED) data, the author proposed that the direction of the nanoplate stacking was along the *c*-axis, perpendicular to the *a*- and *b*-axes of the plates (Fig. 2.15). Although the detailed mechanism were not revealed, it is highly possible that the crystal growth of gadolinium oxide plays a significant role in stacking to form nanoplates.

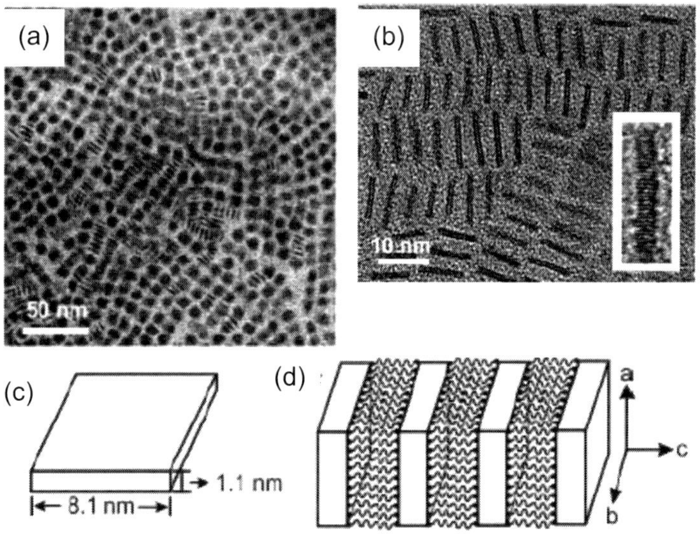

Figure 2.15 (a) TEM images, (b) HRTEM image; inset: cross-fringe image of Gd_2O_3 nanoplates. (c) Proposed model for the nanoplates, (d) assembly of nanoplate stacks. The *c*-axis of cubic Gd_2O_3 crystals is assigned as the thickness direction of the nanoplates [78].

Using similar synthetic route, Si *et al.* [36, 37] reported a range of rare earth oxides nanoplate and nanodisks (Fig. 2.16). The presence of capping ligands (oleylamine and octadecylamine) and, hence, the growth kinetics play a crucial role in forming different phases and morphologies of nanomaterials. It was proposed that possible interparticle interaction among the oleic acid capped nanocrystals and adsorption of liganding solvents onto specific crystal facets are the main driving forces of the morphology evolution of rare earth oxides (Fig. 2.16c).

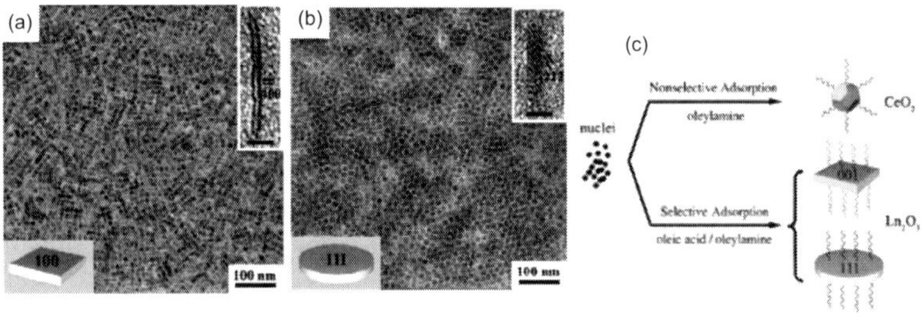

Figure 2.16 TEM images of (a) Pr_2O_3 nanoplates, (b) Eu_2O_3 nanodisks [37], (c) proposed formation mechanism based on selective adsorption of ligands [36].

Two-dimensional rare earth fluorides (REF_3) have been prepared by Yan *et al.* using thermolysis [79, 80] or aqueous solution [81] methods. Composition of the mixed solvents, reaction time, and temperatures were carefully controlled to obtain controlled 2D nanostructure of REF_3 of different sizes and shapes. The authors suggested that the shapes of the rare earth fluorides were also dependent on the cationic radii and phase of the nanocrystals. For example, light REF_3 in a trigonal phase form nanoparticles in trigonal, truncated trigonal, and hexagonal geometry (Fig. 2.17a), which are enclosed by [001] top faces and [110] side faces. Yet, small cations such as Sm^{3+} favors hexagonal shapes, while large cations like La^{3+} favors trigonal shapes. Furthermore, a longer reaction time leads to hexagonal shapes, which evolve from trigonal shapes. Heavy rare earth fluorides in an orthorhombic phase exhibited parallelogram-shaped plates (Fig. 2.17b). The *c*-axis in the trigonal phase corresponds to the *b*-axis in the orthorhombic phase (Fig. 2.17c). Oleic acid attached on the facet of the nanocrystals controls the direction of growth depending on which facet they are adsorbed onto (Fig. 2.17c) [79, 80]. Yet, by tuning the experimental conditions such as addition of more liganding solvents (i.e., oleylamine) or reflux to higher temperature, monodispersity could be controlled [82].

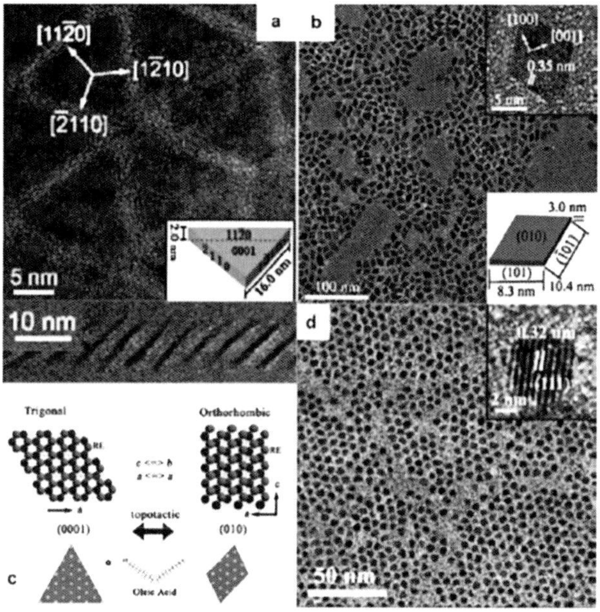

Figure 2.17 (a) LaF_3 nanoplates [79], (b) GdF_3 nanoplates [80], (c) schematic illustration of relation between trigonal and orthorhombic phase of REF_3 and their nanoplates [80], (d) LaOF:Eu nanoparticles.

2.3.4 Core–Shell Rare Earth Nanostructures

The introduction of a shell around a nanoparticle serves multiple purposes: (i) to passivate the surface of luminescent nanoparticles [83, 84], (ii) to provide platform for the fabrication of multifunctional nanocomposites [85–88], (iii) to render them biocompatible [89, 90], (iv) to act as drug carrier [86, 87], and (v) for diagnosis and therapy purposes [86, 91]. In this section, some examples of core-shell rare earth nanoparticles will be discussed.

The idea of coating a shell on luminescent rare earth core nanostructures has been presented by many studies. The shell serves to passivate the luminescent ions of the core, especially those near the surface, from nonradiative decay caused by surface defects as well as from vibrational deactivation from solvents or surface-bound ligands in case of colloidal dispersions. For example, Chow *et al.* [73] reported 30 times luminescence enhancement of ~8 nm $NaYF_4$:Yb,Tm nanocrystals coated with a 1.5 nm thick $NaYF_4$ shell, while Yan *et al.* [92] reported a two-fold increase in luminescence of $NaYF_4$:Yb, Er nanocrystals after growing $NaYF_4$ shell. Very recently, Capobianco *et al.* [83] developed a similar strategy to improve the luminescence of rare earth nanoparticles by fabricating an active/inactive core–shell architecture. By putting an $NaGdF_4$ shell (active: rare earth ions doped; inactive: rare earth ions undoped) around the luminescent core $NaGdF_4$:Yb, Er, a very significant enhancement of luminescence can be obtained. Figure 2.18 shows schematic of the core–shell structure, TEM image, and corresponding enhancement of upconversion efficiency of core–shell nanoparticles. The enhancement of upconversion was attributed to the energy transfer of excited Yb^{3+} ions in the active-shell to the dopant ions in the active-core, which are spatially separated, thereby limiting the effect of concentration quenching observed for core nanoparticles only.

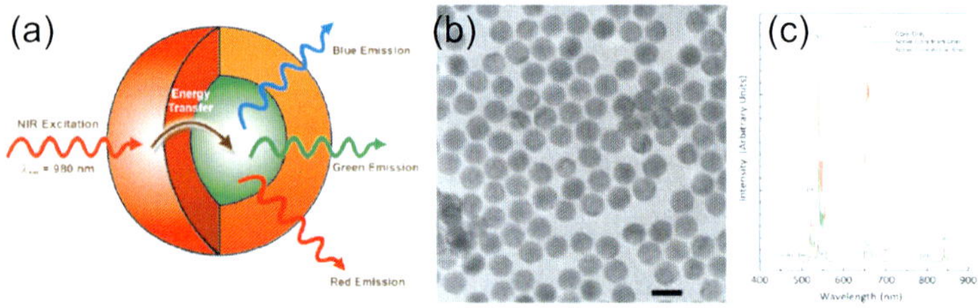

Figure 2.18 (a) An active-core/active-shell nanoparticle architecture showing the absorption of NIR light by the Yb^{3+} rich shell (represented in red) and subsequent energy transfer to the Er^{3+}/Yb^{3+} co-doped core (represented in green), which leads to upconverted blue, green, and red emissions; (b) TEM image of $NaGdF_4:Er^{3+}$, $Yb^{3+}/NaGdF_4:Yb^{3+}$ (active-core/active-shell) nanoparticles; (c) upconversion luminescence spectra of colloidal $NaGdF_4$ [83].

Core–shell strategy has been introduced to induce multifunctionality in nanocrystals. Lu *et al.* [85] reported a core–shell multifunctional nanocomposite of iron oxide-$NaYF_4$:Yb, Er materials (Fig. 2.19a) by a co-precipitation method. The nanocomposites were further functionalized with glutaraldehyde and used in streptavidin immobilization. In another report, Gai *et al.* [87] used a sol-gel method to fabricate a magnetic-luminescent core–shell Fe_3O_4@ SiO_2@$NaYF_4$:Yb^{3+}, Er^{3+}/Tm^{3+} nanocomposite. The mesoporous, upconversion luminescent and magnetic nanocomposites (Fig. 2.19b) act as a multifunctional drug carrier system. Zhang *et al.* [93] reported hydrothermally synthesized Fe_2O_3 and Y_2O_3:Tb nanoparticles embedded in a SiO_2 shell to fabricate multifunctional nanocomposite for potential simultaneous diagnosis and therapy (Fig. 2.19c).

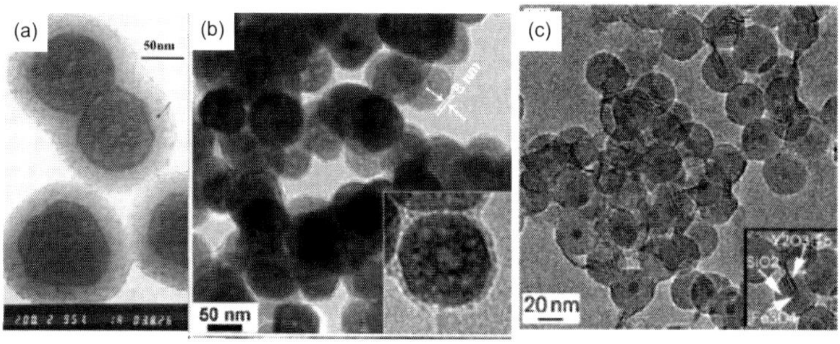

Figure 2.19 TEM image of (a) upconversion fluorescent ($NaYF_4$)-magnetic nanoparticles (iron oxide) with surface-coated silica shells [85]; (b) multifunctional Fe_3O_4@nSiO_2@mSiO_2@$NaYF_4$: Yb^{3+}, Er^{3+} nanocomposties [87]; (c) Fe_3O_4@silica/Y_2O_3:Tb hybrid multifunctional nanostructures [93].

In many cases, highly monodisperse and uniform nanocrystals are synthesized in organic solvents. To use them in biological study, it is essential to render them water-dispersible. Coating these nanoparticles with polymer and/or silica shell renders the nanoparticles water dispersible and provides a further platform for bioconjugation. Das *et al.* [84] demonstrated a strategy of putting a thin layer of silica onto Y_2O_3 nanomaterials to render them water dispersible (Fig. 2.20a), while in another study, Zhong *et al.* [94] demonstrated a hydrothermal method to synthesize PVP/ silica-coated water dispersible particles to be used as biolabels (Fig. 2.20b).

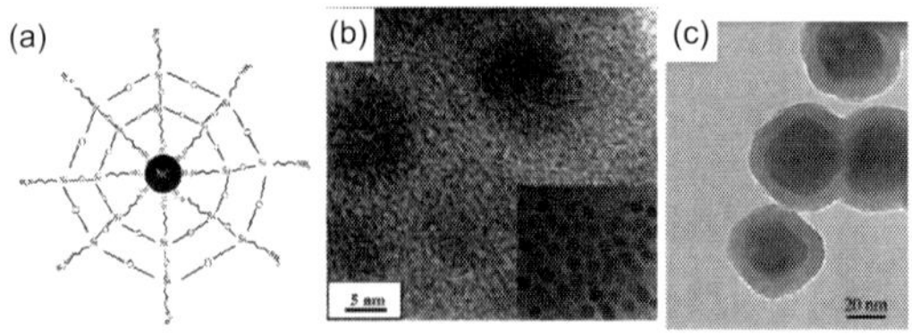

Figure 2.20 (a) Proposed silanization model of nanoparticles with exposed amine functional groups; (b) TEM image of amine functionalized Tb-doped γ-Fe_2O_3 nanocrystals; inset: well-dispersed amine functionalized nanocrystals in water [84]; (b) TEM images of silica-coated PVP/$NaYF_4$:Yb, Er nanocrystals [94].

2.3.5 Hollow Rare Earth Nanostructures

Fabrication of hollow structures has drawn considerable interests due to a wide range of potential applications, including drug and gene delivery, photonics, catalysis, cosmetics, hydrogen production and storage, and as rechargeable batteries [95]. The large void fraction in hollow structures has been successfully used to encapsulate and control the release of sensitive materials such as drugs, cosmetics, and DNA. Methods developed for the fabrication of hollow structures for a range of materials include chemical vapor deposition or epitaxy [96], layer-by-layer technique [97], sacrificed templated or template engaged replacement reaction [98, 99], microemulsion [100], and polymer/surfactant micellar templating [101].

Rare earth materials as hollow nanoparticles have been reported in the form of spherical, fullerene-like, and nanotube structures. Li *et al.* [102, 103] synthesized a number of rare earth hydroxide and fluoride fullerene-like nanoparticles containing single or multiple cavities (RE = Y, Yb, Tm, Er, Ho, Dy, Tb, Gd, Eu, Sm, Nd, Pr, La). Figure 2.21 shows some of the representative TEM images of the different rare earths hollow nanospheres and nanotubes. By controlling the reaction temperature and pH of the solution, they could selectively synthesize hydroxide nanowires, nanosheets, and nanotubes. Again, dehydration, sulfuration, and fluoridation of the hydroxide nanotubes resulted in oxide, oxysulfide, and oxyfluoride nanowires, nanosheets, and nanotubes. TEM image of an yttrium oxysulfide is shown in Fig. 2.21d. In another study, Ma *et al.* [104] reported PrF_3hollow nanosphere by microwave hydrothermal process (Fig. 2.22a). These hollow nanospheres were proposed as drug carrier, or as lubricant additives in catalysis.

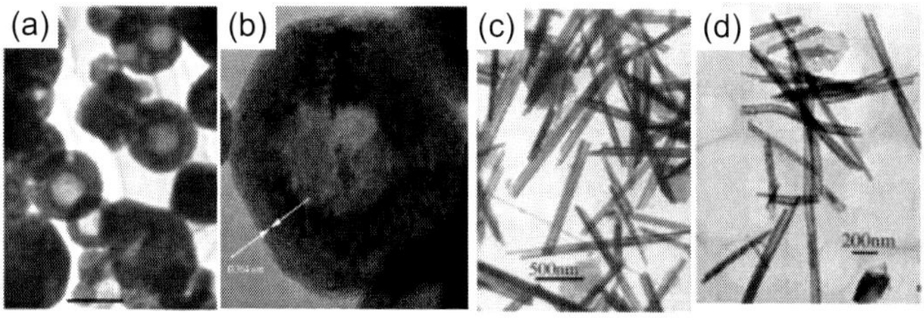

Figure 2.21 (a, b) LaF_3 fullerene-like hollow nanoparticles; HRTEM image of an individual $Eu(OH)_3$ fullerene-like hollow nanoparticles nanoparticle with diameter around 45 nm [102]. (c, d) TEM image of $Yb(OH)_3$ nanotubes and Y_2O_2S nanotubes [103].

Among the earlier works of rare earth nanotubes, Yada *et al.* [105] reported a generalized soft template mediated strategy using sodium dodecylsulfate as surfactant to synthesize rare earth hydroxide and oxide nanotubes. Rare earth oxides nanotubes with inner diameter of 3 nm and wall thickness of 1 nm were produced after calcinations (Fig. 2.22b). Other templates such as cetyltrimethylammonium bromide (CTAB), methyl methacrylate (MMA), polyethylene glycol (PEG), and sodium dodecyl benzenesulfonate (SDBS) have also been used to produce rare earth oxide nanotubes [106–109]. Carbon has also been reported to be used as template in a solvothermal synthesis by Yang *et al.* [110] to obtain Eu_2O_3 nanotubes. Li *et al.* produced a range of rare earth hydroxide nanotubes with and without surfactants at a temperature range of 80–180°C [102, 103].

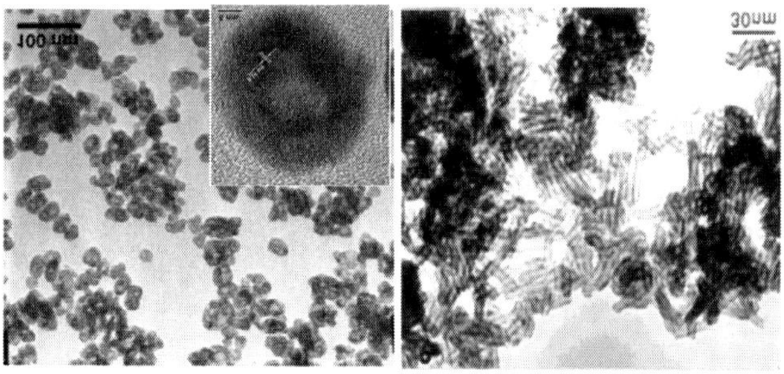

Figure 2.22 (a) TEM image PrF_3 samples prepared by microwave irradiation and hydrothermal treatment of corresponding PrF_3 colloidal precipitate precursor; inset: HRTEM image of an individual as-prepared PrF_3 hollow nanoparticles [104]. (b) TEM images of the Er_2O_3 nanotube templated by dodecylsulfate assemblies [105].

Template or surfactant free rare earth hydroxides have been reported by Xu *et al.* [111] in a hydrothermal synthesis to produce $Dy(OH)_3$ and $Tb(OH)_3$ hollow nanorods at a temperature range of 120–160°C (Fig. 2.23). In general, low temperature and high pH, or high temperature with low pH, are favorable for forming rare earth hydroxide nanotubes. In addition, reported nanotube growth is mostly in a [001] or [100] direction. Rare earth hydroxide nanowires were also reported to grow along [001] direction. In an interesting simulation based study, Martin *et al.* [112] investigated why nanotubes are relatively difficult to synthesize. Using ceria as a model system and adopting a Born model, it was proposed that factors such as polycrystalline behavior of the boundaries, junctions, and dislocations are critical in accommodating the strain in a multilayered wrapped ceria nanotube (Fig. 2.23c).

Figure 2.23 SEM images of: (a) $Dy(OH)_3$, (b) Dy_2O_3 nanotubes [111]. (c) Multilayer wrapped nanotube model showing polycrystalline behavior. The figure indicates important structural features, including boundaries, junctions, and dislocations [112].

$NaREF_4$ upconversion luminescent nanomaterials have also been reported as hollow nanostructures. Zhao *et al.* [113] reported synthesis of multicolor upconversion luminescent nanotubes in an oleic acid assisted hydrothermal method (Fig. 2.24). The method allows the tuning of nanostructures of different morphologies, including nanorods and flower-patterned nanodisks. In a systematic investigation, it was revealed that the amount of fluoride ions (F^-) and oleic acid to alkali ratio (OA/NaOH) played pivotal role in morphology control. For example, the increase in OA/NaOH ratio compare to F^- favors patterned nanodisks formation, while increment of F^- ions favors nanorod formation. The schematic of the formation mechanism is illustrated in Fig. 2.25.

Figure 2.24 (a, b) SEM images of arrays of hexagonal nanotubes, (c) arrays of flower-patterned hexagonal disks of β-$NaYF_4$ [113].

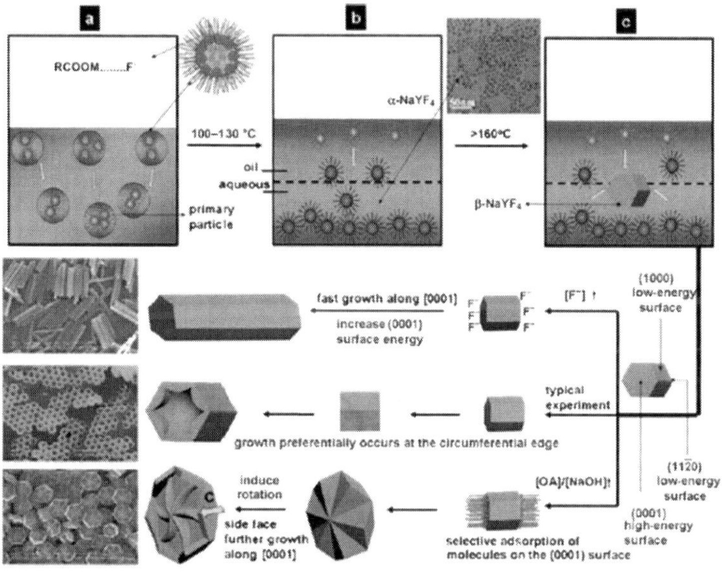

Figure 2.25 Schematic of the formation of arrays of nanocrystals of α-$NaYF_4$ and β-$NaYF_4$ [113].

2.3.6 Complex Rare Earth Nanostructures

In addition to the 0–2D, core–shell, and hollow rare earth nanostructures, some complex nanostructures such as mesostructures, flower-like nanostructures, and superstructures have been reported.

Yan *et al.* [114] reported ceria nanoflower by thermally decomposing $(NH_4)_2Ce(NO_3)_6$ precursor in oleic acid and oleylamine as high boiling solvent at temperature 230–300°C. NO_3^- ions were identified as the conductive species of the solution, which was measured with an electrical resistance measurement probe *in situ*. It was observed that at 230–250°C temperature, sudden increase in free NO^{3-} ions occurred accompanied by violent reaction condition. TEM observation revealed that nanoflowers were formed at that instance. Figure 2.26 shows the flower-like ceria in the form of truncated octahedral structure attached to each other primarily by their [111] facets.

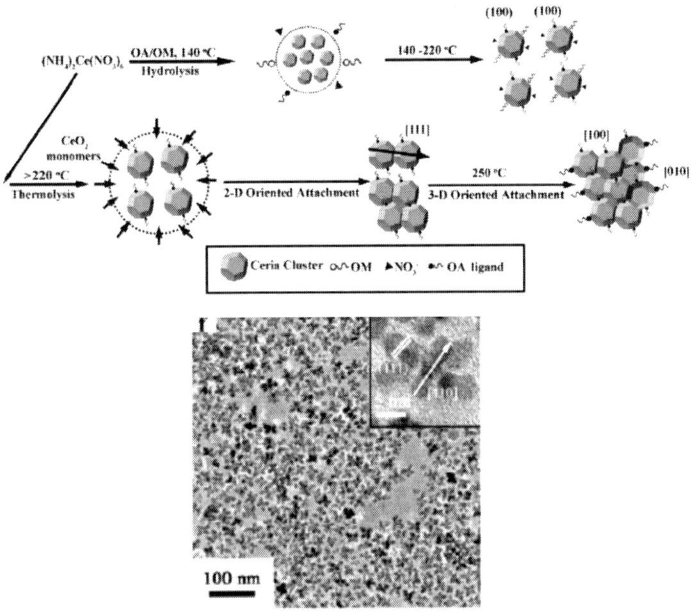

Figure 2.26 (a) TEM image of ceria nanoflowers, (b) schematic of the growth stages of the nanoflowers [114].

Mesostructured materials have attracted much research interest in the area of catalysis. Yan *et al.* [115] developed a sol-gel process combined with evaporation-induced self-assembly in ethanol and used triblock copolymer P123 as a template to fabricate ceria-zirconia mesoporous nanostructures (Fig. 2.27a). The mesoporous ceria–zirconia was tested for catalytic activity with 1% loaded 3 nm Pt nanoparticles for CO oxidation. A temperature-dependent result showed that 50% conversion was obtained at 100°C, while a full 100% conversion was obtained at 350°C (Fig. 2.27b).

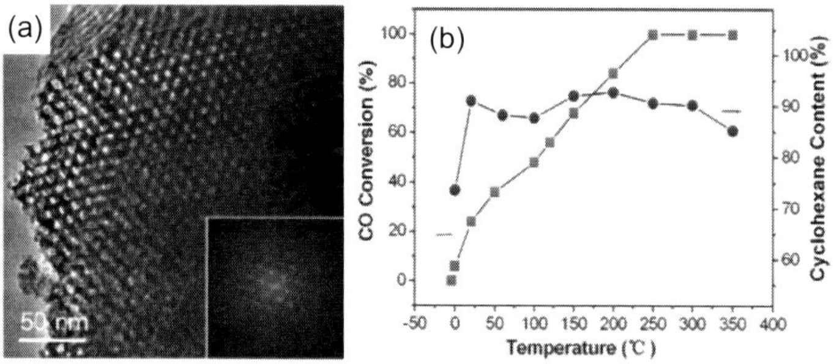

Figure 2.27 (a) TEM images of the mesoporous $Ce_{1-x}Zr_xO_2$ (x = 0.5) nanostructures; inset: corresponding FFT (fast Fourier transform) diffraction image. (b) Catalytic activity of Pt/$Ce_{0.5}Zr_{0.5}O_2$ for CO oxidation and cyclohexene hydrogenation [115].

Superstructures of colloidal nanocrystals capped by surfactants or coated by silica could self-assemble and give rise to novel properties [116]. For rare earth nanostructures, there are a good number of reports on a diverse type of orderly aligned nanostrucutres [37, 79, 80, 117–119]. Some of the examples are shown in Fig. 2.28: (a) $LaPO_4$:Eu polyhedral superlattice, (b) superstructure of SmF_3 nanoplates. The exact formation mechanism of superlattice is not completely understood. However, assembly driven by steric effects for the nanoparticles covered with exchangeable ligands is proposed for many cases.

Figure 2.28 (a) $LaPO_4$:Eu nanopolyhedra superlatice [118], (b) superstructure of SmF_3 nanoplates [80].

Other unusual structures have also been reported. For example, during a shape-evolution experiment of ceria nanoparticles with a relatively large amount of oleylamine, Hyeon *et al.* [120] reported tadpole-shaped nanowires. Tadpole-shaped nanowires consist of a spherical head with a diameter of 3.5 nm and a wire-shaped tail with a diameter of 1.2 nm and length of 27 nm. The low-magnification TEM image and HRTEM images of the tadpole nanostructures are shown in Fig. 2.29a. In a separate study, Li *et al.* [52] came across rice-like YF_3 nanoparticles (Fig. 2.29b) composed of short nanorods and aggregation of nanocrystals. The authors attributed the formation of rice-like nanoparticles to selective adsorption of surfactant molecules in certain planes, which induce further tendency to grow along that direction. For example, surfactant linoleate induces growth along <111> direction for YF_3 nanoparticles, which further agglomerate to form rice-like structure.

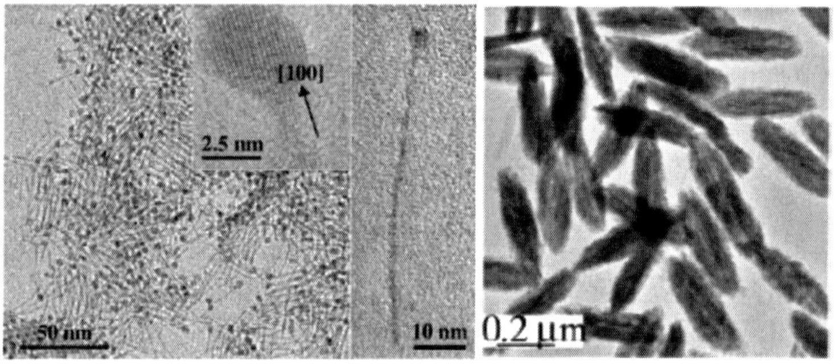

Figure 2.29 (a) TEM images of tadpole-shaped ceria nanocrystals; insets: HRTEM images of a tadpole-shaped nanocrystal [120]. (b) YF_3 rich-like nanocrystals [52].

2.3.7 Phase Control of Rare Earth Nanostructures

Control of the nanocrystalline phase is one of the fundamental ways to tune morphology. Crystal phases of the nanostructures not only influence shape and morphology of the nanocrystals but also modulate the characteristics of the materials [40].

$NaREF_4$ has been investigated as one of the most efficient upconversion hosts, and it exists in two polymorphs: cubic α-$NaREF_4$ and hexagonal β-$NaREF_4$. Between this two, β-phase $NaREF_4$ induces high efficiency upconversion emission. Mai *et al.* [92] reported an oleic acid/oleylamine mediated thermal decomposition method to synthesize cubic α-$NaREF_4$ and hexagonal β-$NaREF_4$ nanocrystals. It is revealed that short reaction time, relatively low temperature, and low Na/RE precursor ratio favor the formation of α-phase nanocrystals. In contrast, β-phase nanocrystals were formed in more rigorous conditions like high temperature, longer reaction time, and high Na/RE ratio. However, phase transformation can take place by addition of extra precursor during the synthesis process. For example, in the above-mentioned experiment, the authors injected additional $Na(CF_3COO)$ precursor into α-$NaREF_4$ nanocrystals, which resulted in β-$NaREF_4$ with longer time. The transformation can also take place in direct synthesis route but is dependent on the type of lanthanides [121].

In a very recent report, Liu *et al.* [40] reported an interesting study to control phase and size simultaneously by doping of lanthanide ions. Generally, doping has been reported to stabilize special crystal phase, modifying electronic properties, modulating magnetism, and tuning emission properties. However, the authors reported that doping can modulate simultaneously crystal phase, size, and emission properties in their work. By doping with lanthanide ions, $NaYF_4$ nanocrystals can be tuned in size, phase (cubic and hexagonal), and upconversion emission wavelengths (green to blue). Figure 2.30 shows the cubic and hexagonal phases of $NaREF_4$ structures and general trend of phase transition from cubic to hexagonal as a function of ionic radius of lanthanide-doped ions. The authors also demonstrated that size and shape can be controlled by doping $NaYF_4$:Yb/Er with Gd^{3+} ions at different concentrations (0–60%). The TEM images show particles of distinct morphology and phase. The authors attributed the size evolution of $NaYF_4$:Yb/Er nanocrystals to the strong effect of the Gd^{3+} dopant ions on crystal growth rate through surface charge modification. DFT calculation suggests that the electron charge density of the crystal surface increases after a Gd^{3+} ion substitutes the Y^{3+} ion in the crystal lattice. The change of electron charge density on the surface of the small-sized nanocrystals can substantially slow the diffusion of negatively charged F^- ions to the surface, because of an increase in charge repulsion, resulting in a tunable reduction of the $NaYF_4$ nanocrystal size.

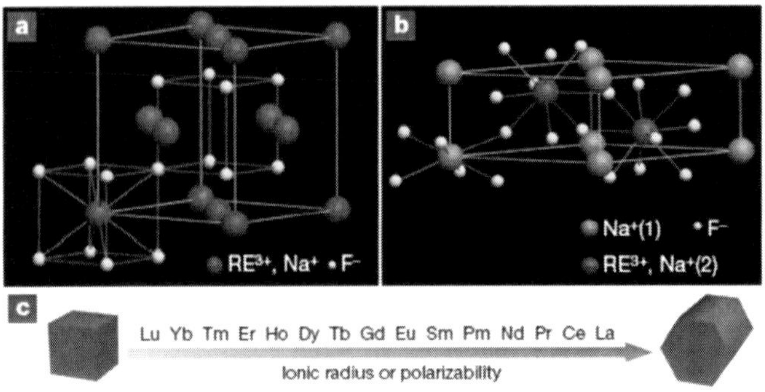

Figure 2.30 (a) Cubic- and (b) hexagonal-phase $NaREF_4$. In the cubic phase, equal numbers of F^- cubes contain cations and vacancies. In the hexagonal phase, an ordered array of F^- ions offers two types of cation sites: one occupied by Na^+ and the other occupied randomly by RE^{3+} and Na^+.

2.4 Modification of the Surface of Rare Earth Nanostructures

The surface phenomena at nanoscale materials are very pronounced compare to their bulk counterparts due to their higher surface-to-volume ratio. Surface of nanoscale materials usually receives two kinds of treatment: (i) surface passivation and (ii) surface functionalization. One of the major applications of rare earth nanomaterials is its potential as fluorescent probe in the bioimaging field. For rare earth doped fluorescent nanomaterials, the presence of any surface dopant ions in incomplete coordination may quench the luminescence. Quenching may take place due to high energy oscillators arising from weakly bound surface impurities, ligands, and solvent due to lack of effective protection by the host lattice. Thus, introduction of an inert crystalline shell around the doped nanoparticles can passivate the surface defect and improve luminescent efficiency. The impact of the passivating surface has been discussed in Section 2.3.4 with the example of luminescence increment of doped rare earth fluorides.

Besides surface passivation, nanocrystals intended for bioapplications often require surface functionalization with different ligands to render them biocompatible. Nanocrystals prepared by high temperature routes, such as thermal decomposition, lack functional moieties for dispersion in aqueous media. Hydrophilic ligands are commonly used to functionalize the nanocrystals surface prior to dispersing in aqueous media or attachment with biomolecules. Different strategies have been explored to provide nanocrystal with properties such as water dispersity and biomolecule conjugation for biomedical applications and biodetection schemes. Table 2.1 shows some of the major strategies reported in the literature, including ligand exchange, ligand oxidation, ligand attraction, layer by layer assembly, and surface silanization.

Ligand exchange has been demonstrated by Yi *et al.*[72] for upconverting $NaYF_4$ nanocrystals where they used polyethylene glycol 600 diacid (HOOC-PEG-COOH) to replace the surface amine ligand (oleylamine) with carboxyl functional groups. The carboxyl functional groups on the nanoparticle surface render them water dispersible. Chen *et al.* [122] developed a ligand oxidation technique in which Lemieux-von Rudloff reagent was used to oxidize the surface oleic acid to azelaic acid. The oxidized ligands have carboxyl groups on the surface, which provide them water dispersity. The same group later reported a strategy based on epoxidation of the double bond followed by reaction with PEG-amines to render the nanocrystals water dispersible [123]. A limitation of this process is that it can only be applied to ligands containing unsaturated carbon–carbon bonds. A ligand attraction process was reported by Yi *et al.* [73] where they coat the core–shell nanocrystals with 25% octylamine and 40% isopropylamine modified poly(acrylicacid) (PAA). Wang *et al.* [124] reported layer-by-layer assembly technique, which is based on the adsorption of alternatively charged polyions on the nanocrystals surface. They reported sequential adsorption of poly(allylamine hydrochloride) and negatively charged poly(sodium 4-styrenesulfonate) onto the $NaYF_4$ surface to generate water-dispersible nanocrystals.

In comparison, the surface silanization technique is versatile and applicable to both hydrophobic and hydrophilic nanocrystals. The Stöber method [125] is a popular silanization technique for hydrophilic nanoparticles, while reverse-microemulsion technique has been adopted for silanization of hydrophobic nanoparticles. The silanization process is based on the hydrolysis and polycondensation of silica precursor (e.g., tetraethoxysilane, TEOS) in the presence of alkali hydroxide (e.g., ammonium hydroxide, NH_4OH). In the surface silanization process, organosilanes with amine functional groups are assembled onto nanoparticles surface, making the nanoparticles water dispersible and biocompatible. This method is attractive because silica coating is very much biocompatible and can be conjugated to a wide range of biomolecules conveniently [126]. Some example of silanization include silica coating of PVP-stabilized $NaYF_4$:Yb,Er [127], a thin layer of silica formation on Fe_2O_3-CdSe magnetic quantum dots [128], and encapsulation of magnetic and fluorescent nanocrystals in silica shell [129, 130].

Table 2.1 Some of the generic strategies for solubilization and functionalization of nanoparticles

Type of strategies	Scheme of strategies	Representative reagents
Ligand exchange [72]		
Ligand oxidation [122]	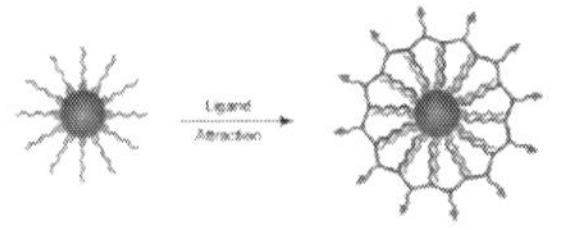	
Ligand attraction [73]	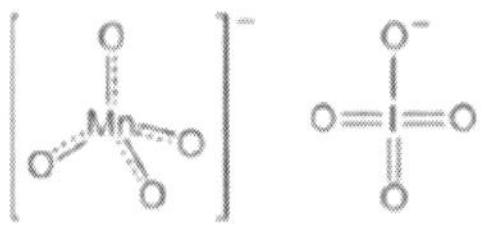	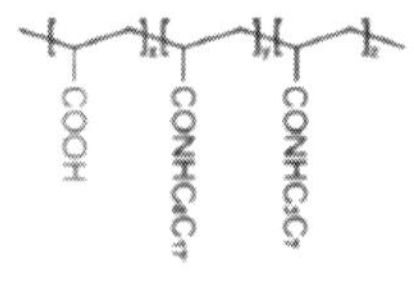
Layer-by-layer assembly [124]	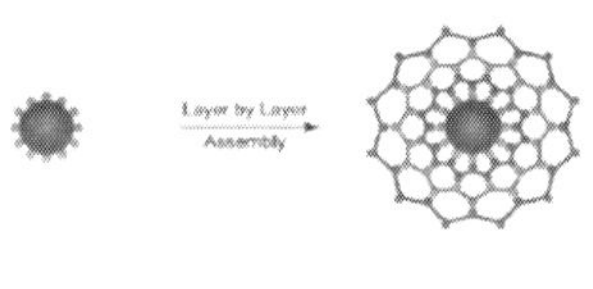	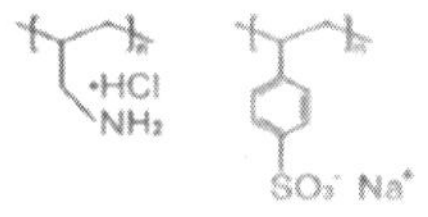
Surface silanization [94]	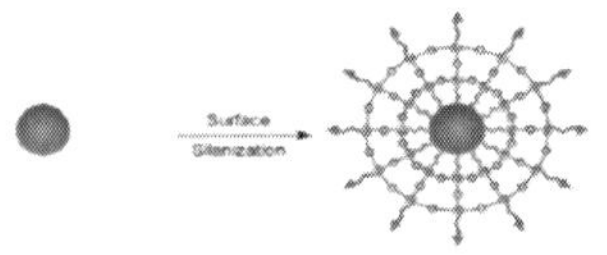	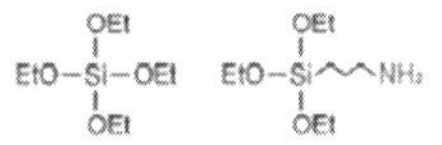

2.5 Conclusion

In this chapter, recent developments in shape-controlled nanostructures together with relevant protocols, working hypothesis, and guiding principles have been presented. It is hoped that these state of the arts can serve as valuable resource, allowing the fabrication of rare earth nanocrystals with specific morphologies suitable in the areas of electronics, photonics, catalysis, information storage, sensing, imaging, and biomedical research and applications.

Morphology control of nanostructures and nanocrystals may initially seem to have attracted widespread interests for fundamental studies, but its goal has gone far beyond their aesthetic appeal. Shape and surface of a nanocrystal determine not only its intrinsic physical and chemical properties but also its relevance to electronic, optical, catalytic, and magnetic applications.

However, the mechanism of the shape-controlled materials synthesis is not well explored till date. The mechanism of size and shape evolution should be elucidated for the design and planning of nanostructures synthesis strategies. The development of *in situ* monitoring techniques such as optical and electrical measurements is, therefore, important, providing opportunities to reveal the underlying mechanisms in crystal nucleation and growth process. Another way to achieve better understanding of nanocrystals growth process is via molecular dynamics simulation. The simulation of crystal growth behavior, surface effect induced phase, morphology, and assembly of rare earth nanocrystals is essential for the understanding of structural characteristics and morphology evolution mechanisms.

Current shape-control strategies are highly dependent on experimental trial-and-error approaches rather than systematic synthetic strategies. The development of more versatile and reliable yet simple synthetic schemes is crucial for tailoring the architecture of nanocrystals with desired components. The knowledge gained will enable further fine tuning of the size, shape, and surfaces of nanocrystals. Another aspect for nanostructure fabrication is to develop strategies to assemble novel nanocrystals of various morphologies into nanodevices, which will be important in future technologies such as fabrication of integrated nanodevices.

References

1. Xia, Y.; Xiong, Y.; Byungkwon, L.; Sara , E. S. (2009) *Angewandte Chemie International Edition*, **48**(1), 60–103.
2. Jun, Y.-W.; Choi, J.-S.; Cheon, J. (2006) *Angewandte Chemie International Edition* **45**(21), 3414–3439.
3. Manna, L.; Scher, E. C.; Alivisatos, A. P. (2000) *Journal of the American Chemical Society* **122**(51), 12700–12706.
4. Andrea, R. T.; Susan, H.; Peidong, Y. (2008) *Small* **4**(3), 310–325.
5. Zhang, J. Z. (2010) *The Journal of Physical Chemistry Letters* **1**(4), 686–695.
6. Taylor-Pashow, K M. L.; Della Rocca, J.; Huxford, R. C.; Lin, W. (2010) *Chemical Communications* **46**(32), 5832–5849.
7. Hyeon, T. (2003) *Chemical Communications* (8), 927–934.
8. Sun, S. (2006) *Advanced Materials* **18**(4), 393–403.
9. Alivisatos, A. P. (1996) *Science* **271**(5251), 933–937.
10. Murray, C. B.; Kagan, C. R.; Bawendi, M. G. (2000) *Annual Review of Materials* **30**, 545–610.
11. Markovich, G.; Collier, C. P.; Henrichs, S. E.; Remacle, F.; Levine, R. D.; Heath, J. R. (1999) *Accounts of Chemical Research* **32**(5), 415–423.
12. Maheshwari, V.; Kane, J.; Saraf, R. (2008) *Advanced Materials* **20**(2), 284–287.
13. Eliseeva, S. V.; Bunzli, J.-C. G. (2010) *Chemical Society Reviews*, **39**, 189–227.
14. Bottrill, M.; Kwok, L.; Long, N. J. (2006) *Chemical Society Reviews*, **35**, 557–571.
15. Bunzli, J.-C. G.; Piguet, C. (2005) *Chemical Society Reviews*, **34**, 1048–1077.
16. Caravan, P. (2006) *Chemical Society Reviews*, **35**, 512–523.
17. Wang, F.; Liu, X. (2009) *Chemical Society Reviews*, **38**, 976–989.
18. Dosev, D.; Nichkova, M.; Kennedy, I. M. (2008) *Journal of Nanoscience and Nanotechnology*, **8**, 1052–1067.
19. Kenyon, A. J. (2002) *Progress in Quantum Electronics*, **26**, 225–284.
20. Shen, J.; Sun, L.-D.; Yan, C.-H. (2008) *Dalton Transactions*, **42**, 5687–5697.
21. Wang, F.; Xue, X.; Liu, X. (2008) *Angewandte Chemie International Edition*, **47**, 906–909.

22. Kim, J.; Piao, Y.; Hyeon, T. (2009) *Chemical Society Reviews*, **38**, 372–390.

23. Das, G. K.; Chan, P. P. Y.; Teo, A.; Loo, J. S. C.; Anderson, J. M.; Tan, T. T. Y. (2010) *Journal of Biomedical Materials Research Part A*, **93A**, 337–346.

24. Auzel, F. (2004) *Chemical Reviews*, **104**, 139–174.

25. Das, G. K.; Heng, B. C.; Ng, S.-C.; White, T.; Loo, J. S. C.; D'Silva, L.; *et al.* (2010) *Langmuir*, **26**, 8959–8965.

26. Abdul Jalil, R.; Zhang, Y. (2008) *Biomaterials*, **29**, 4122–4128.

27. Boyer, J.-C.; Cuccia, L. A.; Capobianco, J. A. (2007) *Nano Letters*, **7**, 847–852.

28. Stephan, H.; Olaf, L.; Markus, H.; Hans-Ulrich, G. (2003) *Angewandte Chemie International Edition*, **42**, 3179–3182.

29. Weissleder, R.; Ntziachristos, V. (2003) *Nature Medicine*, **9**, 123–128.

30. Weissleder, R. (2001) *Nature Biotechnology*, **19**, 316–317.

31. An-Hui, L.; Salabas, E. L.; Ferdi, S. (2007) *Angewandte Chemie International Edition*, **46**, 1222–1244.

32. Xia, Y.; Yang, P.; Sun, Y.; Wu, Y.; Mayers, B.; Gates, B.; *et al.* (2003) *Advanced Materials*, **15**, 353–389.

33. Lin, Y.-S.; Wu, S.-H.; Hung, Y.; Chou, Y.-H.; Chang, C.; Lin, M.-L.; *et al.* (2006) *Chemistry of Materials*, **18**, 5170–5172.

34. Dosev, D.; Guo, B.; Kennedy, I. M. (2006) *Journal of Aerosol Science*, **37**, 402–412.

35. Camenzind, A.; Strobel, R.; Pratsinis, S. E. (2005) *Chemical Physics Letters*, **415**, 193–197.

36. Si, R.; Zhang, Y.-W.; You, L.-P.; Yan, C.-H. (2005) *Angewandte Chemie International Edition*, **44**, 3256–3260.

37. Si, R.; Zhang, Y.-W.; Zhou, H.-P.; Sun, L.-D.; Yan, C.-H. (2006) *Chemistry of Materials*, **19**, 18–27.

38. Zhang, Y.; Guo, J.; White, T.; Tan, T. T. Y.; Xu, R. (2007) *The Journal of Physical Chemistry C*, **111**, 7893–7897.

39. Das, G. K.; Tan, T. T. Y. (2008) *The Journal of Physical Chemistry C*, **112**, 11211–11217.

40. Wang, F.; Han, Y.; Lim, C. S.; Lu, Y.; Wang, J.; Xu, J.; *et al.* (2010) *Nature*, **463**, 1061–1065.

41. Oxtoby, D. W. (2000) *Nature*, **406**, 464–465.

42. LaMer, V. K.; Dinegar, R. H. (1950) *Journal of the American Chemical Society*, **72**, 4847–4854.

43. Watzky, M. A.; Finke, R. G. (1997) *Journal of the American Chemical Society*, **119**, 10382–10400.
44. Colombi Ciacchi, L.; Pompe, W.; De Vita, A. (2001) *Journal of the American Chemical Society*, **123**, 7371–7380.
45. Jin, R.; Cao, Y.; Mirkin, C. A.; Kelly, K. L.; Schatz, G. C.; Zheng, J. G. (2001) *Science*, **294**, 1901–1903.
46. Sun, Y.; Xia, Y. (2002) *Science*, **298**, 2176–2179.
47. Yu, Y. Y.; Chang, S.-S.; Lee, C.-L.; Wang, C. R. C. (1997) *The Journal of Physical Chemistry B*, **101**, 6661–6664.
48. Ahmadi, T. S.; Wang, Z. L.; Green, T. C.; Henglein, A.; El-Sayed, M. A. (1996) *Science*, **272**, 1924–1925.
49. Huynh, W. U.; Dittmer, J. J.; Alivisatos, A. P. (2002) *Science*, **295**, 2425–2427.
50. Cui, Y.; Banin, U.; Björk, M. T.; Alivisatos, A. P. (2005) *Nano Letters*, **5**, 1519–1523.
51. Wang, X.; Zhuang, J.; Peng, Q.; Li, Y. (2005) *Nature*, **437**, 121–124.
52. Wang, X.; Zhuang, J.; Peng, Q.; Li, Y. (2006) *Inorganic Chemistry*, **45**, 6661–6665.
53. Lemyre, J.-L.; Ritcey, A. M. (2005) *Chemistry of Materials*, **17**, 3040–3043.
54. Arriagada, F. J.; Osseo-Asare, K. (1999) *Journal of Colloid and Interface Science*, **211**, 210–220.
55. Murray, E. P.; Tsai, T.; Barnett, S. A. (1999) *Nature*, **400**, 649–651.
56. Corma, A.; Atienzar, P.; Garcia, H.; Chane-Ching, J.-Y. (2004) *Natural Materials*, **3**, 394–397.
57. Fu, Q.; Saltsburg, H.; Flytzani-Stephanopoulos, M. (2003) *Science*, **301**, 935–938.
58. Terribile, D.; Trovarelli, A.; de Leitenburg, C.; Primavera, A.; Dolcetti, G. (1999) *Catalysis Today*, **47**, 133–140.
59. Morshed, A. H.; Moussa, M. E.; Bedair, S. M.; Leonard, R.; Liu, S. X.; El-Masry, N. (1997) *Applied Physics Letters*, **70**, 1647–1649.
60. Yang, S.; Gao, L. (2006) *Journal of the American Chemical Society*, **128**, 9330–9331.
61. Yan, L.; Yu, R.; Chen, J.; Xing, X. (2008) *Crystal Growth and Design*, **8**, 1474–1477.
62. Cho, K.-S.; Talapin, D. V.; Gaschler, W.; Murray, C. B. (2005) *Journal of the American Chemical Society*, **127**, 7140–7147.

63. Yan, Z.-G.; Yan, C.-H. (2008) *Journal of Materials Chemistry*, **18**, 5046–5059.
64. Wang, X.; Li, Y. (2002) *Angewandte Chemie International Edition*, **41**, 4790–4793.
65. Fang, Y.-P.; Xu, A.-W.; Song, R.-Q.; Zhang, H.-X.; You, L.-P.; Yu, J. C.; *et al.* (2003) *Journal of the American Chemical Society*, **125**, 16025–16034.
66. Luo, F.; Jia, C.-J.; Song, W.; You, L.-P.; Yan, C.-H. (2004) *Crystal Growth and Design*, **5**, 137–142.
67. Deng, H.; Yang, S.; Xiao, S.; Gong, H.-M.; Wang, Q.-Q. (2008) *Journal of the American Chemical Society*, **130**, 2032–2040.
68. Liu, J.; Li, Y. (2007) *Advanced Materials*, **19**, 1118–1122.
69. Stouwdam, J. W.; Raudsepp, M.; van Veggel, F. C. J. M. (2005) *Langmuir*, **21**, 7003–7008.
70. Buissette, V.; Giaume, D.; Gacoin, T.; Boilot, J.-P. (2006) *Journal of Materials Chemistry*, **16**, 529–539.
71. Riwotzki, K.; Haase, M. (1998) *The Journal of Physical Chemistry B*, **102**, 10129–10135.
72. Yi, G. S.; Chow, G. M. (2006) *Advanced Functional Materials*, **16**, 2324–2329.
73. Yi, G.-S.; Chow, G.-M. (2007) *Chemistry of Materials*, **19**, 341–343.
74. Wang, L.; Li, Y. (2006) *Nano Letters*, **6**, 1645–1649.
75. Mai, H.-X.; Zhang, Y.-W.; Si, R.; Yan, Z.-G.; Sun, L.-d.; You, L.-P.; *et al.* (2006) *Journal of the American Chemical Society*, **128**, 6426–6436.
76. Puntes, V. F.; Krishnan, K. M.; Alivisatos, A. P. (2001) *Science*, **291**, 2115–2117.
77. Wang, J.; Liu, Q.; Liu, Q. (2007) *Optical Materials*, **29**, 593–597.
78. Cao, Y. C. (2004) *Journal of the American Chemical Society*, **126**, 7456–7457.
79. Zhang, Y.-W.; Sun, X.; Si, R.; You, L.-P.; Yan, C.-H. (2005) *Journal of the American Chemical Society*, **127**, 3260–3261.
80. Sun, X.; Zhang, Y. W.; Du, Y. P.; Yan, Z. G.; Si, R.; You, L. P.; *et al.* (2007) *Chemistry – A European Journal*, **13**, 2320–2332.
81. Stouwdam, J. W.; van Veggel, F. C. J. M. (2002) *Nano Letters*, **2**, 733–737.
82. Du, Y.-P.; Zhang, Y.-W.; Sun, L.-D.; Yan, C.-H. (2007) *The Journal of Physical Chemistry C*, **112**, 405–415.

83. Vetrone, F.; Naccache, R.; Mahalingam, V.; Morgan, C. G.; Capobianco, J. A. (2009) *Advanced Functional Materials*, **19**, 2924–2929.
84. Zhang, Y.; Das, G. K.; Xu, R.; Tan, T. T. Y. (2009) *Journal of Materials Chemistry*, **19**, 3696–3703.
85. Lu, H.; Yi, G.; Zhao, S.; Chen, D.; Guo, L.-H.; Cheng, J. (2004) *Journal of Materials Chemistry*, **14**, 1336–1341.
86. Yang, P.; Quan, Z.; Hou, Z.; Li, C.; Kang, X.; Cheng, Z.; *et al.* (2009) *Biomaterials*, **30**, 4786–4795.
87. Gai, S.; Yang, P.; Li, C.; Wang, W.; Dai, Y.; Niu, N.; *et al.* (2010) *Advanced Functional Materials*, **20**, 1166–1172.
88. Mi, C.; Zhang, J.; Gao, H.; Wu, X.; Wang, M.; Wu, Y.; *et al.* (2010) *Nanoscale*, **2**, 1141–1148.
89. Hifumi, H.; Yamaoka, S.; Tanimoto, A.; Akatsu, T.; Shindo, Y.; Honda, A.; *et al.* (2009) *Journal of Materials Chemistry*, **19**, 6393–6399.
90. Fortin, M.-A.; Petoral, R. M. P. Jr; Söderlind, F.; Klasson, A.; Engström, M.; Veres, T.; *et al.* (2007) *Nanotechnology*, **18**, 395501.
91. Wang, F.; Banerjee, D.; Liu, Y.; Chen, X.; Liu, X. (2010) *Analyst*, **135**, 1839–1854.
92. Mai, H.-X.; Zhang, Y.-W.; Sun, L.-D.; Yan, C.-H. (2007) *The Journal of Physical Chemistry C*, **111**, 13721–13729.
93. Zhang, Y.; Pan, S.; Teng, X.; Luo, Y.; Li, G. (2008) *The Journal of Physical Chemistry C*, **112**, 9623–9626.
94. Li, Z.; Zhang, Y. (2006) *Angewandte Chemie International Edition*, **45**, 7732–7735.
95. Lou, X. W.; Archer, L. A.; Yang, Z. (2008) *Advanced Materials*, **20**, 3987–4019.
96. Goldberger, J.; He, R.; Zhang, Y.; Lee, S.; Yan, H.; Choi, H.-J.; *et al.* (2003) *Nature*, **422**, 599–602.
97. Caruso, F.; Caruso, R. A.; Mohwald, H. (1998) *Science*, **282**, 1111–1114.
98. Sun, Y.; Mayers, B. T.; Xia, Y. (2002) *Nano Letters*, **2**, 481–485.
99. Peng, Q.; Dong, Y.; Li, Y. (2003) *Angewandte Chemie International Edition*, **42**, 3027–3030.
100. Schacht, S.; Huo, Q.; Voigt-Martin, I. G.; Stucky, G. D.; Schuth, F. (1996) *Science*, **273**, 768–771.
101. Yu, J.; Guo, H.; Davis, S.; Mann, S. (2006) *Advanced Functional Materials*, **16**, 2035–2041.

102. Wang, X.; Li, Y. (2003) *Angewandte Chemie International Edition*, **42**, 3497–3500.

103. Wang, X.; Li, Y. (2003) *Chemistry – A European Journal*, **9**, 5627–5635.

104. Ma, L.; Chen, W.-X.; Zheng, Y.-F.; Zhao, J.; Xu, Z. (2007) *Materials Letters*, **61**, 2765–2768.

105. Yada, M.; Mihara, M.; Mouri, S.; Kuroki, M.; Kijima, T. (2002) *Advanced Materials*, **14**, 309–313.

106. Tang, C.; Bando, Y.; Golberg, D.; Ma, R. (2005) *Angewandte Chemie International Edition*, **44**, 576–579.

107. Guo, C.; Cao, M.; Hu, C. (2005) *Journal of Nanoscience and Nanotechnology*, **5**, 184–187.

108. Shi, W.; Yu, J.; Wang, H.; Yang, J.; Zhang, H. (2006) *Journal of Nanoscience and Nanotechnology*, **6**, 2515–2519.

109. Li, W.; Wang, X.; Li, Y. (2004) *Chemical Communications*, **2**, 164–165.

110. Yang, H.; Zhang, D.; Shi, L.; Fang, J. (2008) *Acta Materialia*, **56**, 955–967.

111. Xu, A.-W.; Fang, Y.-P.; You, L.-P.; Liu, H.-Q. (2003) *Journal of the American Chemical Society*, **125**, 1494–1495.

112. Martin, P.; Parker, S. C.; Sayle, D. C.; Watson, G. W. (2007) *Nano Letters*, **7**, 543–546.

113. Zhang, F.; Wan, Y.; Yu, T.; Shi, Y.; Xie, S.; Li, Y.; *et al.* (2007) *Angewandte Chemie International Edition*, **46**, 7976–7979.

114. Zhou, H. P.; Zhang, Y. W.; Mai, H. X.; Sun, X.; Liu, Q.; Song, W. G.; *et al.* (2008) *Chemistry – A European Journal*, **14**, 3380–3390.

115. Yuan, Q.; Liu, Q.; Song, W.-G.; Feng, W.; Pu, W.-L.; Sun, L.-D.; *et al.* (2007) *Journal of the American Chemical Society*, **129**, 6698–6699.

116. Redl, F. X.; Cho, K. S.; Murray, C. B.; O'Brien, S. (2003) *Nature*, **423**, 968–971.

117. Huo, Z.; Chen, C.; Chu, D.; Li, H.; Li, Y. (2007) *Chemistry – A European Journal*, **13**, 7708–7714.

118. Mai, H.-X.; Zhang, Y.-W.; Sun, L.-D.; Yan, C.-H. (2007) *Chemistry of Materials*, **19**, 4514–4522.

119. Zhao, F.; Yuan, M.; Zhang, W.; Gao, S. (2006) *Journal of the American Chemical Society*, **128**, 11758–11759.

120. Yu, T.; Joo, J.; Park, Y. I.; Hyeon, T. (2005) *Angewandte Chemie International Edition*, **44**, 7411–7414.

121. Aebischer, A.; Heer, S.; Biner, D.; Krämer, K.; Haase, M.; Güdel, H. U. (2005) *Chemical Physics Letters*, **407**, 124–128.

122. Chen, Z.; Chen, H.; Hu, H.; Yu, M.; Li, F.; Zhang, Q.; *et al.* (2008) *Journal of the American Chemical Society*, **130**, 3023–3029.

123. Hu, H.; Yu, M.; Li, F.; Chen, Z.; Gao, X.; Xiong, L.; *et al.* (2008) *Chemistry of Materials*, **20**, 7003–7009.

124. Wang, L.; Ruoxue, Y.; Ziyang, H.; Lun, W.; Jinghui, Z.; Jie, B.; *et al.* (2005) *Angewandte Chemie International Edition*, **44**, 6054–6057.

125. Stöber, W.; Fink, A.; Bohn, E. (1968) *Journal of Colloid and Interface Science*, **26**, 62–69.

126. Burns, A.; Ow, H.; Wiesner, U. (2006) *Chemical Society Reviews*, **35**, 1028–1042.

127. Zhengquan, L.; Yong, Z. (2006) *Angewandte Chemie International Edition*, **45**, 7732–7735.

128. Selvan, S. T.; Patra, P. K.; Ang, C. Y.; Ying, J. Y. (2007) *Angewandte Chemie International Edition*, **46**, 2448–2452.

129. Liu, Z.; Yi, G.; Zhang, H.; Ding, J.; Zhang, Y.; Xue, J. (2008) *Chemical Communications*, 694–696.

130. Yi, D. K.; Selvan, S. T.; Lee, S. S.; Papaefthymiou, G. C.; Kundaliya, D.; Ying, J. Y. (2005) *Journal of the American Chemical Society*, **127**, 4990–4991.

Chapter 3

Rare Earth Nanomaterials in Fluorescence Microscopy

Muthu Kumara Gnanasammandhan[a] and Yong Zhang[b]
[a] *Division of Bioengineering, National University of Singapore, 117576, Singapore*
[b] *Nanoscience and Nanotechnology Initiative, National University of Singapore, 117576, Singapore*
biezy@nus.edu.sg

3.1 Introduction and Overview

The past decade has seen the development of a plethora of nanoparticles covering a wide range of applications. Nanomaterials have significantly different properties than bulk materials and have various advantages like high surface area to volume ratio, and their size plays a crucial role in applications where they are used. This chapter focuses on the application of rare earth nanomaterials in one of the most important biological applications, namely fluorescence microscopy.

The discovery of microscope in the late 1500's was one of the major milestones in the field of biology, and in the last few decades, there has been an exponential increase in the microscopic techniques available, which have greater resolution and capabilities of tracking

Rare Earth Nanotechnology
Edited by Timothy T.Y. Tan

ISBN 978-981-4316-30-9 (Hardcover), 978-981-4364-20-1 (eBook)
www.panstanford.com

individual molecules and other processes inside cells. For applications where visualization of different cellular components is necessary or when there is a need for detection of trace amount of molecules, it was thought that labeling the molecule of interest with fluorescent dyes will address this need, and hence fluorescence microscopy came into play. Fluorescence microscope excites fluorescent molecules used to label cells known as fluorophores with a particular wavelength and detects the fluorescence, which is of another wavelength.

Initially, organic dyes such as fluorescein isothiocyanate (FITC), rhodamine, and coumarin were used for fluorescence labeling due to its various advantageous properties such as small size, ease of attachment to biomolecules, and high quantum yield. However, they suffer from limitations such as low photostability, which hinders long-term monitoring, and also overlap of excitation and emission spectra for some organic dyes. The emergence of fluorescent tags of biological origin such as green fluorescent protein (GFP) and DsRed provides the advantages of themselves being able to genetically fuse to the protein of interest so that the protein can be tracked in real time. But they too face some shortcomings such as low quantum yield and low fluorescence lifetime. Recently, many types of fluorescent nanoparticles were developed to overcome the limitations of the earlier fluorescent probes. A main class is polymeric nanoparticles, where organic dyes were encapsulated in polymers to increase stability and amplify signals. However, they were not stable enough for long-term studies. The next class comprises fluorescent semiconductor crystals, also known as quantum dots, which have gained tremendous popularity. Their color output can be tuned by changing their sizes, and they have good photostability with narrow emission spectrum. But they too face some problems such as photoblinking and high cytotoxicity. Studies are under way to overcome these problems. The final class of nanoparticles consists of rare earth fluorides, which are believed to have many superior properties compared to their predecessors.

3.2 Rare Earth Nanoparticles

Rare earth elements constitute a class of elements found in the 6th and 7th periods of the periodic table along with lanthanum and actinium. It is divided into two main series known as the lanthanide and actinide series. The elements in the lanthanide series have gained attention over the past few years due to their unique optical characteristics, especially in the area of biological applications. Most of the lanthanides exist in a trivalent state (except for cerium and europium), and their unique optical properties arise from electronic transitions within the 4f shell or from 4f–5d shell. Many synthesized rare earth nanomaterials consisted of one rare earth element, or combinations of more than one rare earth elements, to produce a wide range of spectral characteristics. They can be tuned to emit from the UV to the near-infrared (NIR) region by the process of upconversion or downconversion depending on the composition of nanoparticles. Downconversion is a process of absorbing a high-energy photon and emitting low-energy photons. On the contrary, upconversion is a process in which two or more low-energy photons are absorbed to produce a single high-energy photon. The mechanism of upconversion and downconversion is explained in Fig. 3.1.

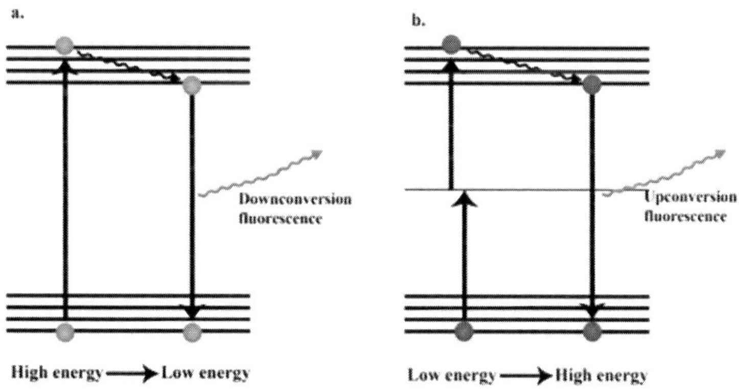

Figure 3.1 Mechanism of downconversion and upconversion fluorescence.

For biological applications and especially for optical microscopy, the process of upconversion is highly advantageous as these nanoparticles can absorb in the NIR region and produce visible light, eliminating background autofluorescence as biological components do not absorb in the NIR region. NIR light has good tissue penetration compared to UV and visible light. Therefore, rare earth upconversion nanoparticles are excellent candidates for *in vivo* imaging. There are different types of rare earth nanoparticles with various composition and they are discussed in the following sections.

3.2.1 Lanthanide Chelates

Lanthanide ions by themselves do not exhibit fluorescence as they have weak absorbance. But when they are chelated with some chelating agents such as β-diketonates or EDTA, they absorb in the UV region and produce strong fluorescence in the visible region. The ligand absorbs and transfers the energy to lanthanide ions, which undergo excitation and produce fluorescence. Chelating the lanthanide ions also protects them from the quenching effect. Chelates of lanthanides such as Eu^{3+} and Tb^{3+} fluoresce brightly, Nd^{3+} and Tm^{3+} have weak fluorescence, whereas Gd^{3+} and Lu^{3+} do not fluoresce at all. The structures of Eu^{3+} and Tb^{3+} chelates are shown in Fig. 3.2.

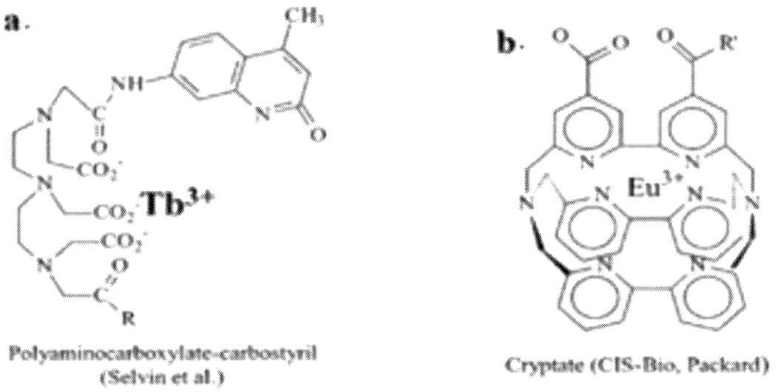

Figure 3.2 Structure of Eu^{3+} and Tb^{3+} lanthanide chelates. Adapted from [1], reprinted with permission, from the *Annual Review of Biophysics and Biomolecular Structure*, Volume 31 © 2002 by Annual Reviews www.annualreviews.org.

Different lanthanide chelates have different excitation and emission wavelengths, and their spectral properties also vary, which makes them good candidates for multiplex labeling applications. They exhibit large Stokes shift, high quantum yield, long fluorescence lifetime, and sharp emission peaks, which are very advantageous for a biological label [2]. Usually, lanthanide chelates of europium, samarium, and dysprosium are used as labeling agents as they exhibit the most efficient energy transfer from the chelate and are being used in routine microscopic techniques and especially for time-resolved fluorescence microscopy [3]. They are also available commercially and marketed under the trade name RadiantDOTS and TorrentDOTS etc., by Biopal™.

3.2.2 Lanthanide-Doped Nanoparticles

They are another class of rare earth nanomaterials which has found a lot of biological applications compared to lanthanide chelates and a lot of research is being done currently to synthesize and biofunctionalize these nanoparticles and use them for various applications. These nanoparticles consist of a host crystal lattice doped with lanthanide ions in low concentrations. The host material should have a crystal lattice, which can fit the dopant ions so that the ions are held tightly in the crystal. The host material is usually inorganic in nature with low phonon energy and should provide an environment to sensitize the fluorescence of the dopant ions. Inorganic oxides, oxysulfides, bromides, chlorides, and fluorides have been investigated as host material and doped with various lanthanide ions, synthesized in various sizes and shapes. A perspective by Shen *et al.* explains in detail the different types of rare earth nanoparticles synthesized so far for biological applications [4]. The schematic in Fig. 3.3 shows the model of lanthanide-doped nanoparticles with trivalent lanthanide ions doped in the host crystal lattice.

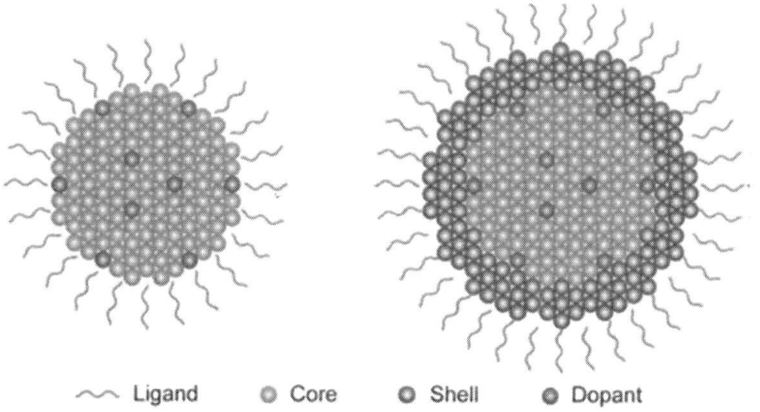

Figure 3.3 Schematic showing the model of lanthanide-doped nanoparticles. Adapted from [5], reprinted with permission from the Royal Society of Chemistry, Copyright © 2009.

Lanthanide-doped nanoparticles show several advantageous properties such as long fluorescence lifetime (in the order of milliseconds), high quantum yield, sharp emission peaks, color tuning depending on the ions doped, and good resistance to photobleaching from environmental and other factors. The extreme stability of lanthanides is attributed to the electronic transitions occurring in the ions, as compared to the involvement of chemical bonds in organic fluorophores and fluorescent proteins, which suffer from low photostability. Lanthanide-doped nanoparticles can be divided into two main groups, namely, downconversion and upconversion nanoparticles depending on the mechanism by which they fluoresce.

3.2.2.1 Downconversion nanoparticles

This is a class of lanthanide-doped nanoparticles that fluoresce by a mechanism similar to conventional organic fluorophores and quantum dots. These nanoparticles absorb high-energy photons and emit low-energy photons, i.e., they absorb in the UV region of the electromagnetic spectrum and emit in the visible and infrared region. These nanomaterials also exhibit another phenomenon known as quantum cutting whereby a single high-energy photon is absorbed to produce two photons of lower energy. Even though there are only a few lanthanides, various factors such as the combinations in which they are used, the phonon energy of the host crystal lattice, the oxidation state of the lanthanide ions, and the concentration ratio of dopant ions to the host play a major role in the production of these nanoparticles to give multiple emission peaks.

Lanthanide-doped downconversion nanoparticles have been synthesized from a variety of inorganic host materials such as $LaPO_4$, LaF_3, $NaYF_4$, and $CePO_4$ and doped with various lanthanide ions such as Eu^{3+}, Yb^{3+}, Tb^{3+}, Er^{3+}, and Nd^{3+} [6–8]. Sivakumar *et al.* have also developed LaF_3 nanoparticles doped with Nd^{3+}, which can absorb in the visible region at 514 nm and emit in the NIR region at 1070 nm [9]. Although a wide variety of these nanoparticles have been developed, they are not promising for biological applications due to its inherent disadvantages similar to other biological probes. They have been mainly used in the electronics industry for lamps and optical amplifiers and also for the construction of lasers [10].

3.2.2.2 Upconversion nanoparticles

Fluorescent lanthanide-doped nanoparticles, which possess upconversion characteristics, have gained popularity as a class of fluorescent labels recently. Upconversion occurs when two or more photons of lower energy are absorbed, followed by the emission of a photon of higher energy. Upconversion mechanisms include energy transfer upconversion, excited state absorption, and photon avalanche [11]. The phenomenon of upconversion was first noticed in the 1960s in bulk materials. Only in the past decade, there has been a lot of focus on developing lanthanide upconversion nanomaterials. A review by Feng *et al.* provides a clear picture on the mechanisms of upconversion and also gives the selection criteria for the host material and lanthanide dopants, which are necessary for the synthesis of lanthanide upconversion nanomaterials [5].

Table 3.1 Comparison of different classes of fluorescent nanoparticles

	Organic dye-doped nanoparticles	Quantum dots (QDs)	Upconversion nanoparticles (UCNs)
Size	50–500 nm	2–10 nm	50–200 nm
Autofluorescence	High	High	Low
Light penetration depth	Medium/High	Medium/High	High
Photodamage	Medium/Low	Medium/Low	Low
Cytotoxicity	Medium	High/Medium	Low
Biocompatibility	Good	Good	Good
Photostability	Low	High	High
Excitation wavelength	UV/Vis/NIR	UV/NIR	NIR
Cost	Low	High	Low
Excitation Radiation toxicity	Medium/Low	Medium/Low	Low
Multiplexing assays	N/A	Good	Good

Note: Adapted from [12], reprinted with permission from the Royal Society Publishing, © 2010.

The presence of multiple metastable states in lanthanide ions makes them good candidates for upconversion. Initially, these nanomaterials were being used in solid state lasers and in other optical applications. But their unique characteristic of absorbing NIR light and emitting in the visible region paved way for their exploration in biological applications. Upconversion is very well suited for biological applications because NIR light does not excite any tissue components, and hence there is almost zero background autofluorescence, which is especially advantageous in microscopic techniques [13]. The use of NIR light for exciting the nanoparticles also offers high tissue penetration depth compared to UV and visible light, enabling high resolution *in vivo* imaging [14]. The phototoxicity from NIR is negligible. The advantages of upconversion nanoparticles compared to other fluorescent nanoparticles such as organic dye-doped nanoparticles and quantum dots are highlighted in Table 3.1. Upconversion nanoparticles show excellent photostability, chemical stability, and thermal stability. Good photostability ensures long-term tracking of these nanoparticles, which is not possible in conventional organic fluorescent labels, which photobleach very quickly. They also do not exhibit the phenomenon of "on-off" photoblinking, which is prevalent in quantum dots and causes the loss of information about the process under study when it is in an "off" state [15]. These particles can also be tuned to emit various colors depending on the type of crystal lattice, lanthanide dopant ions, and their respective concentration [16]. Figure 3.4 shows upconversion $NaYF_4$ nanoparticles doped with different concentrations of Yb^{3+}, Er^{3+}, and Tm^{3+}. These multicolored nanoparticles have potential application in multiplex biological imaging, where different cellular components can be labeled with different colors and imaged.

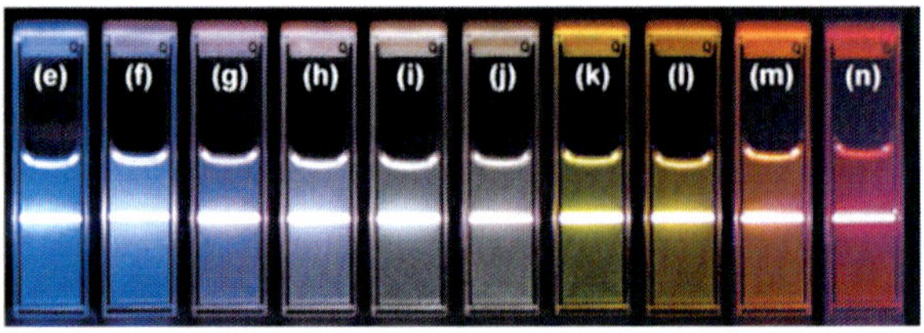

Figure 3.4 Upconversion $NaYF_4$ nanoparticles doped with different concentrations of Yb^{3+}, Er^{3+}, and Tm^{3+} showing multicolor emissions. Adapted from [16], reprinted with permission from the American Chemical Society, Copyright © 2008.

Lanthanide-doped upconversion nanoparticles have been synthesized from a variety of inorganic host crystals and lanthanide ions. However, certain host crystals such as LaF_3, $YbPO_4$, Y_2O_3, and $LuPO_4$ are found to be more efficient, while $NaYF_4$ was found to be the most efficient host lattice for upconversion [17]. The commonly used lanthanide ions for doping are Yb^{3+}, Tm^{3+}, Er^{3+}, and Ho^{3+}. The nanoparticles synthesized from these materials absorb usually around 980 nm in the NIR range and has multiple emission peaks from 350 to 800 nm.

3.3 Nanoparticle Synthesis

There are a variety of methods available for the synthesis of rare earth-based nanoparticles. In this chapter, we will focus only on the synthesis of rare earth upconversion nanoparticles, which is being used in fluorescence microscopy applications. Early synthesis methods employed very high temperature, and the size was in the range of a few hundred nanometers [18]. Further optimization of these methods paved way for the synthesis of upconversion nanoparticles with controlled size, good fluorescence intensity, tunable emission wavelengths, controllable morphologies, and with good stability [16, 19].

The basic principle underlying the synthesis of upconversion nanoparticles is that the precursors of the host crystal and lanthanide salts are mixed in various concentrations, usually in a solvent and the system is subsequently heated. Crystal nuclei emerge, which then grow to form the nanoparticles. Surfactants such as EDTA, CTAB, or other ligands such as polyvinyl pyrrolidone, polyethylene imine are used to control particle growth [22]. Synthesis methods that do not use high temperatures (such as the thermal decomposition method) use higher pressure with lower temperature, rendering the synthesis process quicker [23].

The size, shape, and phase of the crystals are modified by controlled variation of various reaction parameters such as temperature, time, and reactant concentration (Fig. 3.5). For example, for the synthesis of $NaYF_4$ nanocrystals, Zhang *et al.* demonstrated that increasing the concentration of precursor (NaF) yielded nanorods of β-$NaYF_4$ crystals, while modifying the time and temperature of the reaction yielded mixture of cubic α-$NaYF_4$ and hexagonal β-$NaYF_4$ crystals at 230°C for 1 h. Extended heating time resulted in hexagonal nanotubes and β-$NaYF_4$. When the oleic acid/NaOH concentration was varied, flower-patterned β-$NaYF_4$ crystals were observed [21]. Thus, different methods and different parameters have to be considered to obtain $NaYF_4$ nanocrystals of different crystalline phases and morphologies. The most common methods employed for the synthesis of these nanoparticles are tabulated in Table 3.2. Each method has its own advantages and disadvantages. But the synthesis of these nanoparticles has become much easier and user friendly than the methods initially developed. Recently, Zhenquan *et al.* have developed an user-friendly method for the synthesis of hexagonal phase $NaYF_4$ nanocrystals with very high upconversion fluorescence [24].

For more detailed discussion on rare earth nanomaterials synthesis, please refer to chapter 1.

Table 3.2 Summary of different synthesis methods available

Method	Example (hosts)	Remarks
Co-precipitation	LaF_3 $NaYF_4$ $LuPO_4$ $YbPO_4$	Fast growth rate without the need for costly equipment and costly procedures. Post-heat treatment typically required
Thermal decomposition	LaF_3 $NaYF_4$ GdOF	Expensive air-sensitive metal precursors. High quality monodisperse nanocrystals. Toxic byproducts.
Hydro (solvo) thermal synthesis	LaF_3 $NaYF_4$ $La_2(MoO_4)_3$ YVO_4	Cheap raw materials. No post-heat treatment. Excellent control over particle size and shape. Specialized reaction vessels required.
Sol-gel processing	ZrO_4 TiO_2 $BaTiO_3$ $Lu_3Ga_5O_3$ YVO_4	Cheap raw materials. Calcinations at high temperatures required.
Combustion synthesis	Y_2O_3 Gd_2O_3 La_2O_2S	Time and energy saving. Considerable particle aggregation.
Flame synthesis	Y_2O_3	Time saving and readily available

Note: Adapted from [5], reprinted with permission from the Royal Society of Chemistry, Copyright © 2009)

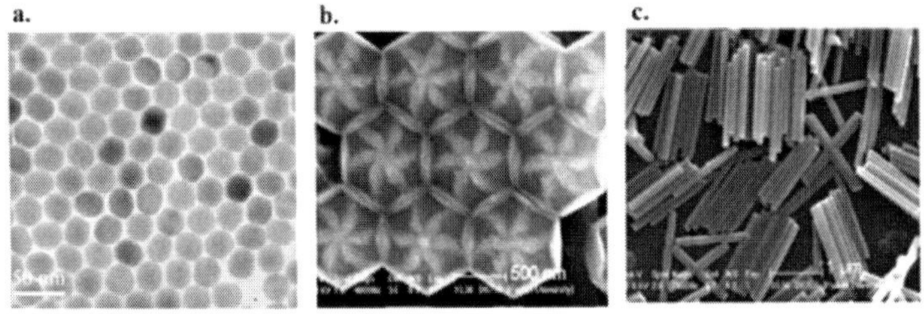

Figure 3.5 Scanning electron micrographs showing different shapes of lanthanide-doped nanoparticles (a) [20] (b–c) [21]. Reprinted with permission, Copyright © 2007, 2009, WILEY-VCH Verlag GmbH & Co. KGaA, Weinheim.

3.4 Surface Functionalization

The nanoparticles synthesized by most of the aforementioned methods cannot be used directly for biological applications. They are typically synthesized in an organic environment at high temperatures without any ligand capping and, therefore, the nanoparticles are hydrophobic. It is very important that they are dispersible in aqueous solutions for use in biological applications. Various strategies have been undertaken to coat the surface of nanoparticles with hydrophilic ligands or other inert materials to make them water soluble. Another essential process is the attachment of functional groups on the surface of nanoparticles, such as the amine and carboxyl group, for further attachment of antibodies, aptamers, and other targeting agents.

Surface functionalization can be done by exchanging hydrophobic ligands on the surface of nanoparticles by a ligand-exchange reaction, where the existing ligands on the surface of nanoparticles, such as oleate, are exchanged with ligands such as polyethylene glycol (PEG) to render them dispersible in water [25]. The ligands can also be modified by a chemical reaction to make them hydrophilic. For example, Zhou *et al.* developed a method, in which a direct oxidation of oleic acid into azelaic acid or azelaic aldehyde was performed, making the nanoparticles water dispersible and easier for further modification [26].

Other methods such as coating amphiphilic polymers on the surface of nanoparticles, where the hydrocarbon backbone of the polymer is attracted to the hydrophobic ligands on the surface of the particles by ligand attraction, and layer-by-layer assembly of oppositely charged polyions on the surface of nanoparticles, have also been developed [27, 28]. Apart from the above methods, coating the surface of nanoparticles with an amorphous layer of silica has gained prominence and is being routinely used. Van Veggel *et al.* coated a layer of silica on nanoparticles via surface silanization [9]. Coating the hydrophobic surface of nanoparticles with silica and coating individual nanoparticles rather than their aggregates is a difficult task. Li *et al.* reported the coating of $NaYF_4$ nanoparticles with a thin layer of silica with tunable thickness and good monodispersity [29]. Recently, a mesoporous layer of silica coating was also performed to facilitate the loading of therapeutic molecules such as drug or gene into the pores of nanoparticles for simultaneous delivery and imaging experiments [30].

Sometimes, the core of nanoparticles is first coated with another layer of host material or an inert crystalline material before functionalization. This is done to passivate the ligands on the surface of the core, to fill the surface defects and to prevent fluorescence quenching by impurities attached on the surface [31, 32]. This helps to enhance the fluorescence of nanoparticles and provides a core-shell structure to nanoparticle (as shown in Fig. 3.3).

Chapter 2 provides a more detailed discussion on rare earth surface modification and functionalization.

3.5 Applications

Rare earth-based nanoparticles have been investigated in a wide range of biological applications ranging from imaging, immunocytochemistry, DNA detection, and the like. This section will provide in-depth discussions of the applications of these nanoparticles in fluorescence microscopy. Fluorescence microscopy involves the visualization of cells or cellular components with subcellular resolution by labeling them with fluorescent probes. Initially, wide-field fluorescence microscopes were used and later laser scanning confocal microscopy was developed, which overcame some of the earlier limitations, followed by two photon microscopy, Total Internal Reflection Microscopy (TIRF), and Photoactivation Localization Microscopy (PALM). All these techniques have their own advantages and limitations. Even though they have very good resolution, some challenges still exist, which cannot be overcome by improving the microscopic technique alone but through the improvement of fluorescent labels.

The use of conventional fluorophores in fluorescent microscopy is limited by various factors such as high background autofluorescence, photobleaching, and phototoxicity, as mentioned earlier. Rare earth-based upconversion fluorescent nanoparticles are able to address all these issues. Various studies have reported the benefits of using these nanoparticles for imaging biological systems, both *in vitro* and *in vivo*, as described in detail in the following section.

3.5.1 *In vitro* Microscopy

Imaging of live cells and their various cellular components and tracking specific biomolecules are essential parts of cell and molecular biology. The routinely used fluorescent lanthanide-doped upconversion labels for imaging are those consisting of a $NaYF_4$ host doped with $Yb^{3+}/Er^{3+}/Tm^{3+}$. They have been found to show the most efficient upconversion of NIR to visible light [17]. The use of an NIR excitation source helps in significantly reducing the background autofluorescence. Various approaches have been attempted to overcome this problem by various algorithms to filter out the background autofluorescence. A NIR excitation does not excite the endogenous components of cells, and tissues do not absorb in this region, hence almost zero background autofluorescence. Comparative studies have shown the advantage of imaging with upconversion nanoparticles over conventional fluorophores with respect to background autofluorescence, which ultimately gives rise to a high signal-to-noise ratio [33].

Nanoparticles of various sizes, with and without targeting agents, have been used for imaging. Li *et al.* have reported the use of silica-coated $NaYF_4$: Yb, Er upconversion nanoparticles for nontargeted imaging of live HT-29 cells. Incubation of the nanoparticles with the cells for 24 h showed good cellular uptake [20]. Targeted imaging helps the nanoparticles to attach to the cell membrane and to be internalized very quickly. Many studies have been done to target various tumor cell lines aimed at developing sensitive diagnostic assays and for receptor-mediated drug delivery to these cells. The same group has also reported the targeted imaging of HT-29 and SK-BR-3 cells by nanoparticles attached with folic acid and anti

Her2 antibody, respectively. Folic acid was used to target the folate receptors of HT-29 cells and anti-Her2 to target the Herceptin receptors of SK-BR-3 cell [34, 35]. The cellular labeling and uptake was imaged using a confocal fluorescence microscope fitted with a 980 nm NIR laser.

Multicolor imaging of HeLa cells using multicolor $NaYbF_4$: Tm^{3+}, Er^{3+}, Ho^{3+} nanoparticles was reported by Wang *et al.* They also showed the specificity of the upconversion nanoparticles conjugated with anti-CEA8 antibody, which targeted CEA8 antigens on HeLa cells, and that the nanoparticles took only an hour to attach and be internalized in the presence of targeting agents [36]. The availability of multicolored nanoparticles helps in multiple labeling of different cellular components with different colors. A similar study was done by Zako *et al.*, but single-colored Y_2O_3: Er^{3+} upconversion nanoparticles, conjugated with cyclic arginine-glycine-aspartate (RGD) peptide, were used for specific imaging of integrin-positive tumor cells [37]. These studies made use of an inverted fluorescence microscope fitted with a 980 nm NIR laser for exciting the nanoparticles.

The fate of the nanoparticles after internalization, their cytotoxicity, biocompatibility, and the time it requires to be cleared from the system are crucial issues to be addressed. A study by Boyer *et al.* employed $NaYF_4$: Yb^{3+}, Tm^{3+} upconversion nanoparticles for NIR imaging of ovarian cancer cells (CaOV3). The intracellular uptake of these nanoparticles was visualized by Transmission Electron Microscopy (TEM). The results showed that the nanoparticles had been internalized and accumulated in the vesicles present in the cytoplasm; none were found inside the nucleus. It also showed that there was no significant aggregation of nanoparticles inside the cells [25]. The cytotoxicity of these nanoparticles, influence of various concentrations of nanoparticles on cell viability, and biodistribution of these nanoparticles in Wistar rats have been studied. The study showed good biocompatibility, and the nanoparticles were cleared from the system within seven days [38].

Apart from the fluorescence imaging of cells, these nanoparticles can also be used to detect nucleic acids, proteins, and other biomolecules by fluorescence microscopy (Fig. 3.6). Rijke *et al.* showed that lanthanide-doped upconversion nanoparticles were better reporter molecules than dyes such as Cy5 for detecting cDNA hybridization in a microarray. The sensitivity of the technique was found to be higher when upconversion fluorescence was detected by a wide-field digital fluorescence microscope in comparison to Cy5 fluorescence detected by a conventional microplate reader [18]. It was also used for studying the delivery and release of siRNA in live cells using upconversion-based Fluorescent Resonance Energy Transfer (FRET). The siRNA was stained with BOBO-3, which was excited by the emission of the upconversion nanoparticles [39]. This upconversion-based FRET offers many advantages over conventional FRET, such as higher sensitivity, nonoverlapping of excitation wavelength, and low autofluorescence. This upconversion-based FRET has also been employed in the detection of proteins such as avidin and trace amounts of glucose [27, 41].

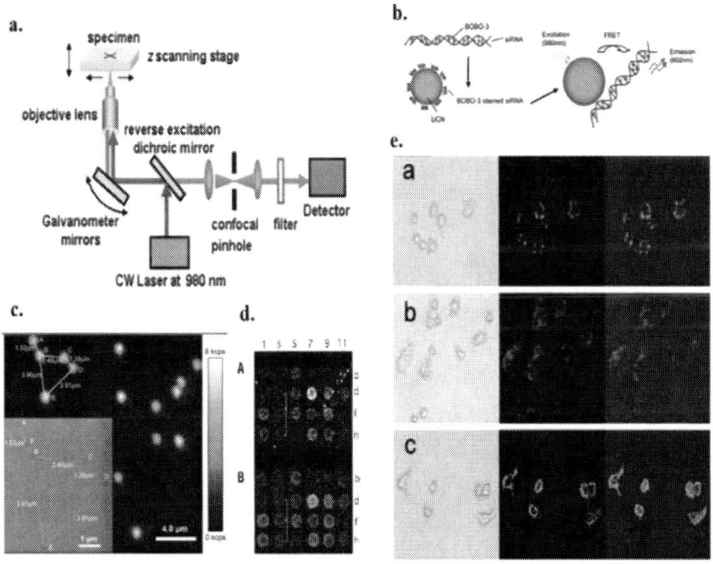

Figure 3.6 *In vitro* microscopy applications using lanthanide nanomaterials. (a) Schematic showing the setup of Laser Scanning Upconversion Luminescence Microscopy (LSUCLM) [13], (b) upconversion Fluorescent Resonance Energy Transfer (FRET) for tracking of siRNA [39], (c) single molecule tracking of lanthanide-based upconversion nanoparticles [40], (d) microarray using lanthanide-based upconversion nanoparticles [18], (e) imaging of cells [36]. Reprinted with permission, Copyright © 2001, 2009, 2010, American Chemical Society, National Academy of Sciences and Nature Publishing Group.

The properties of these lanthanide-based upconversion nanoparticles also make them very good candidates for single molecular imaging and tracking. A recent study has shown the potential of these nanoparticles for single molecular tracking, by analyzing individual nanoparticles in different medium and checking various parameters such as fluorescence intensity, photobleaching, and photoblinking [40]. All the studies mentioned above made use of different types of fluorescence microscopes with slight modifications, such as the addition of a 980 nm NIR source for excitation. There are currently no commercial microscopes specifically designed for upconversion luminescence imaging. Efforts have been taken to develop specialized microscopes for such applications. A new method for upconversion luminescence imaging, known as Laser Scanning Upconversion Luminescence Microscopy (LSUCLM), was developed. It removes out-of-focus signals and increases the resolution of upconversion luminescence images [13]. A schematic layout of a LSUCLM setup is shown in Fig. 3.6a.

3.5.2 *In vivo* Microscopy

The principle behind *in vivo* fluorescence imaging is similar to the normal fluorescence microscopic technique, but it is in a larger scale, thus more challenging. The major obstacles are the thick tissues, which are opaque to visible light and generate a very high background autofluorescence. So for *in vivo* fluorescence imaging, the NIR window (650–900 nm) is utilized for imaging as the biological components have a very low absorption in this region. Initially, NIR dyes such as Cy5.5 and Cy7 were used [42], but they displayed rapid photobleaching, low water solubility, and low detection sensitivity. With the development of nanotechnology, NIR dyes were encapsulated in nanocapsules for imaging. However, lanthanide-doped upconversion nanoparticles show superiority over other NIR labeling agents. Their inherent properties, as discussed earlier, make them very attractive for *in vivo* fluorescence imaging with very high spatial and temporal resolution.

In vivo imaging using upconversion nanoparticles as labels has been performed on various model systems ranging from nematodes to plants and small animals such as mice. Lim *et al.* imaged the digestive tract of the nematode *Caenorhabditis elegans* with Y_2O_3: Yb, Er upconversion nanoparticles. The worms were fed with the nanoparticles, and they were imaged at various time points to track their movement through the digestive tract. They were monitored for 24 h without any photobleaching observed. The upconversion fluorescence from the digestive tract of the worms is shown in Fig. 3.7b [43]. Similarly, $NaYF_4$: Yb, Er upconversion nanoparticles were able to enter *Arabidopsis* and *Phalaenopsis* plants through the roots, and upconversion fluorescence was observed in the roots, shoots, and leaves [44].

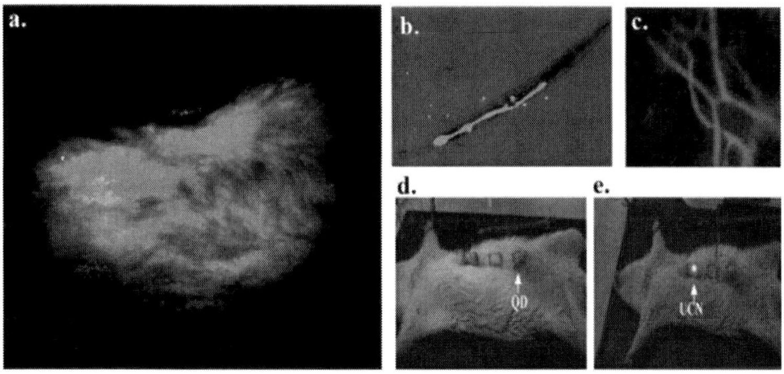

Figure 3.7 *In vivo* microscopy applications using lanthanide-based nanoparticles. (a) *in vivo* NIR imaging of Balb-c mice [45], (b) imaging the digestive tract of *C. elegans* [43], (c) imaging of blood vessels [46], (d, e) comparison of the fluorescence emission of quantum dots and lanthanide-based upconversion nanoparticles [35]. Reprinted with permission, Copyright © 2006, 2007, 2008, 2009, American Chemical Society, Elsevier Ltd., Royal Society of Chemistry.

The effectiveness of these nanoparticles in small animal imaging has also been reported. Chatterjee *et al.* reported the imaging of $NaYF_4$: Yb, Er upconversion nanoparticles, which were subcutaneously injected in various parts of Wistar rats, such as the upper leg and groin. The upconversion fluorescence was also compared with the fluorescence from green-emitting quantum dots, and it was found to have better tissue penetration depths [35]. A comparative study between the effectiveness of Cy5.5 (NIR dye) and upconversion nanoparticles for *in vivo* imaging was also reported using transillumination fluorescence imaging [47]. They were also used for the real time imaging of myoblasts transfected with upconversion nanoparticles in a living mouse model. Time lapse confocal imaging was done to follow the transfected myoblasts in the blood vessels of the mice [14].

NIR imaging by lanthanide-doped upconversion nanoparticles, which can absorb at 980 nm and emit around 800 nm, shows good promise for deeper tissue imaging compared to those described in the earlier section, as both excitation and emission are in the NIR region. Scott *et al.* reported the use of Y_2O_3 upconversion nanoparticles, which were surface modified with a NIR dye, carbocyanin, for multichannel imaging of blood vessels in the ear lobe of mice. It gave an option of using the emission of both upconversion emission and dye emission [46]. Nyk *et al.* reported the use of $NaYF_4$: Yb^{3+}, Tm^{3+} nanoparticles, which have an emission in the NIR region around 800 nm for the whole-body imaging of mice after tail-vein injection of the nanoparticles. Imaging showed the accumulation of nanoparticles in the liver and spleen 2 h post-injection, and the obtained images were in good resolution with spectral unmixing and high contrast [45].

3.5.3 Multimodal Imaging

The field of multimodal imaging is evolving rapidly and involves the combination of one or more imaging modalities for a single examination. It utilizes the advantages of both the imaging modalities to provide better resolution. Imaging techniques such as magnetic resonance imaging (MRI), nuclear medicine and optical imaging are usually used in combinations. Multimodal imaging techniques such as positron emission tomography (PET)– computerized axial tomography (CT) have reached clinical practice. The recent interest in molecular imaging has promoted the use of optical imaging in combination with other imaging modalities.

Since upconversion fluorescence microscopy has many advantages for in-depth molecular imaging, combining this technique with other imaging modalities such as MRI will be very advantageous. So a lot of research is being done in this area, and upconversion nanoparticles with an MRI contrast agent have been developed. Li *et al.* have reported the development of $NaYF_4$ upconversion nanoparticles with a silica coating, incorporated with Gd^{3+} for multimodal imaging. These nanoparticles provided good T_1- and T_2-weighted contrast, and they were functionalized with carboxylic groups on the surface for further attachment of biomolecules for targeting [48]. Similarly, core–shell upconversion nanoparticles were synthesized with $NaGdF_4$: Er^{3+}, $Yb^{3+}/NaGdF_4$, where the Gd^{3+} ions in the host matrix act as an MRI contrast agent [49].

3.5.4 Multifunctional Nanoparticles

Upconversion nanoparticles have been shown to be excellent fluorescent labels. They have also been shown to achieve other functions such as delivery of biomolecules. Nanoparticle-based delivery systems are widely used. They have numerous advantages since these upconversion nanoparticles can be biofunctionalized coated with porous materials, which can be utilized for the delivery of various therapeutic molecules, enabling simultaneous imaging and therapy. This will be very important in the treatment of various diseases.

A few examples showing the multifunctionality of these nanoparticles are elaborated here. Various reports have shown the supremacy of these nanoparticles in photodynamic therapy. Photosensitizers such as zinc phthalocyanine or porphyrin were loaded onto the upconversion nanoparticles and delivered specifically to cancer cells through targeting agents. The photosensitizers produced singlet oxygen upon NIR activation. NIR activation produces upconversion fluorescence, which further activates the photosensitizer and ultimately causes cell death [50–52]. They can be simultaneously used for the detection and imaging of tumors. Multifunctional upconversion nanoparticles were also used for the delivery and tracking of siRNA to cells. The siRNA was attached to the silica coating on the nanoparticles, and the upconversion fluorescence was used to monitor the uptake. Upconversion FRET was used to study the release of siRNA after entering into the cells [34, 39]. Thus these particles show good multifunctionality and can be used especially for simultaneous delivery and imaging.

3.6 Summary and Outlook

This chapter gives an overview of the different types of nanomaterials synthesized from lanthanides, their synthesis methods and surface modifications, which are done before using them for fluorescence imaging. Lanthanide chelates and rare earth downconversion nanoparticles are not very popular because of their inherent limitations of low penetration depth and high phototoxicity, and hence downconversion nanoparticles are finding applications mainly in electronics and optics rather than biology. Significant studies have been done in the last decade for developing these rare earth nanomaterials with upconversion fluorescence as an excellent fluorescent label for different fluorescence microscopy applications. Such upconversion nanomaterials, absorbing in the NIR and emitting in the visible/NIR regions, have significantly contributed to improvement in resolution, sensitivity and decrease in background autofluorescence in fluorescence microscopy techniques. The use of NIR as an excitation source has advantages such as low phototoxicity and high signal-to-noise ratio as NIR does not excite biological samples and, hence, has negligible background autofluorescence. Despite rapid development in this field, there are still many challenges and avenues for improvement. The only advantage of organic dyes over these nanoparticles is their smaller size. So the development of sub-10 nm nanoparticles with good upconversion fluorescence will allow the fluorescence labeling of intracellular components.

Various reports show the promise of lanthanide-doped upconversion nanoparticles in deep tissue *in vivo* fluorescence microscopy. However, there is a long way to go before it reaches clinical testing. More studies are needed to improve the imaging depth and resolution *in vivo*. This is possible with the current development of lanthanide-doped upconversion nanoparticles with the capabilities of multimodal imaging. Syntheses of nanoparticles, which show upconversion fluorescence and MRI contrast, have been reported. But incorporation of isotopes such as ^{124}I or ^{125}I can be done to enable other multimodal imaging techniques such as PET and SPECT in combination with fluorescence imaging using the upconversion fluorescence. All these developments will give a new dimension to *in vitro* and *in vivo* fluorescence molecular imaging. Thus the use of fluorescent rare earth nanomaterials in fluorescence imaging techniques is a burgeoning, interdisciplinary field holding a lot of promise in the years to come.

Acknowledgements

The authors would like to acknowledge the financial support from Singapore A*STAR SBIC and National University of Singapore.

References

1. Selvin, P. R. (2002) Principles and biophysical applications of lanthanide-based probes. *Annu. Rev. Biophys Biomol. Struct.*, **31**(1), 275–302.
2. Bornhop, D. J., *et al.* (1999) Fluorescent tissue site-selective lanthanide chelate, Tb-PCTMB for enhanced imaging of cancer. *Anal. Chem.*, **71**(14), 2607–2615.
3. Väisänen, V., *et al.* (2000) Time-resolved fluorescence imaging for quantitative histochemistry using lanthanide chelates in nanoparticles and conjugated to monoclonal antibodies, *Luminescence*, **15**(6), 389–397.
4. Shen, J., Sun, L.-D. (2008) Luminescent rare earth nanomaterials for bioprobe applications, *Dalton Trans.*, **42**, 5687–5697.
5. Feng Wang, X. L. (2009) Recent advances in the chemistry of lanthanide-doped upconversion nanocrystals, *Chem. Soc. Rev.*, **38**, 976–989.
6. DiMaio, J. R., Kokuoz, B., and Ballato, J. (2006) White light emissions through downconversion of rare earth doped LaF3 nanoparticles, *Opt. Express*, **14**(23), 11412–11417.
7. Kömpe, K., *et al.* (2003) Green-emitting $CePO_4$: Tb/$LaPO_4$ core–shell nanoparticles with 70% photoluminescence quantum yield, *Angew. Chem. Int. Ed.*, **42**(44), 5513–5516.
8. Riwotzki, K., *et al.* (2000) Liquid-phase synthesis of doped nanoparticles: colloids of luminescing LaPO4: Eu and CePO4: Tb particles with a narrow particle size distribution, *J. Phys. Chem. B*, **104**(13), 2824–2828.
9. Sivakumar, S., Diamente, P. R., and van Veggel, F. C. J. M. (2006) Silica-coated Ln^{3+}-doped LaF_3 nanoparticles as robust down- and upconverting biolabels, *Chem. Eur. J.*, **12**(22), 5878–5884.
10. Singh, S. K., Kumar, K., and Rai, S. B. (2009) Er^{3+}/Yb^{3+} codoped Gd_2O_3 nanophosphor for optical thermometry, *Sens. Actuators A*, **149**(1), 16–20.
11. Auzel, F. (2003) Upconversion and anti-Stokes processes with f and d ions in solids, *Chem. Rev.*, **104**(1), 139–174.
12. Jiang, S., Gnanasammandhan, M. K., and Zhang, Y. (2010) Optical imaging-guided cancer therapy with fluorescent nanoparticles, *J. R. Soc. Interface*, **7**(42), 3–18.
13. Yu, M., *et al.* (2009) Laser scanning upconversion luminescence microscopy for imaging cells labeled with rare earth nanophosphors, *Anal. Chem.*, **81**(3), 930–935.

14. Idris, N. M., *et al.* (2009) Tracking transplanted cells in live animal using upconversion fluorescent nanoparticles, *Biomaterials*, **30**(28), 5104–5113.

15. Kuno, M., *et al.* (2001) "On"/"off" fluorescence intermittency of single semiconductor quantum dots, *J. Chem. Phys.*, **115**(2), 1028–1040.

16. Wang, F., and X. Liu (2008) Upconversion multicolor fine-tuning: visible to near-infrared emission from lanthanide-doped $NaYF_4$ nanoparticles, *J. Am. Chem. Soc.*, **130**(17), 5642–5643.

17. Kramer, K. W., *et al.* (2004) Hexagonal sodium yttrium fluoride based green and blue emitting upconversion phosphors, *Chem. Mater.*, **16**(7), 1244–1251.

18. van de Rijke, F., *et al.* (2001) Upconverting phosphor reporters for nucleic acid microarrays, *Nat. Biotech.*, **19**(3), 273–276.

19. Guo, H., and Qiao, Y. M. (2009) Preparation, characterization, and strong upconversion of monodisperse Y_2O_3: Er^{3+}, Yb^{3+} microspheres, *Opt. Mater.*, **31**(4), 583–589.

20. Li, Z. Q., Zhang, Y., and Jiang, S. (2009) Multicolor core/shell-structured upconversion fluorescent nanoparticles, *Adv. Mater.*, **21**(47), 4765–4769.

21. Zhang, F., *et al.* (2007) Uniform nanostructured arrays of sodium rare earth fluorides for highly efficient multicolor upconversion luminescence, *Angew. Chem. Int. Ed.*, **46**(42), 7976–7979.

22. Zeng, J.-H., *et al.* (2005) Synthesis and upconversion luminescence of hexagonal-phase $NaYF_4$: Yb^{3+} phosphors of controlled size and morphology, *Adv. Mater.*, **17**(17), 2119–2123.

23. Zhang, Y.-W., *et al.* (2005) Single-crystalline and monodisperse LaF_3 triangular nanoplates from a single-source precursor, *J. Am. Chem. Soc.*, **127**(10), 3260–3261.

24. Zhengquan, L., and Yong, Z. (2008) An efficient and user-friendly method for the synthesis of hexagonal-phase $NaYF_4$: Yb, Er/Tm nanocrystals with controllable shape and upconversion fluorescence, *Nanotechnology*, **19**(34), 345606.

25. Boyer, J.-C., *et al.* (2009) Surface modification of upconverting $NaYF_4$ nanoparticles with PEG–phosphate ligands for NIR (800 nm) biolabeling within the biological window, *Langmuir*, **26**(2), 1157–1164.

26. Zhou, H.-P., *et al.* (2009) Clean and flexible modification strategy for carboxyl/aldehyde-functionalized upconversion nanoparticles and their optical applications, *Adv. Funct. Mater.*,. **19**(24), 3892–3900.

27. Wang, L., *et al.* (2005) Fluorescence resonant energy transfer biosensor based on upconversion-luminescent nanoparticles13, *Angew. Chem. Int. Ed.*, **44**(37), 6054–6057.

28. Yi, G.-S., and Chow, G.-M. (2006) Water-soluble $NaYF_4$: Yb, Er(Tm)/ $NaYF_4$/polymer core/shell/shell nanoparticles with significant enhancement of upconversion fluorescence, *Chem. Mater.*, **19**(3), 341–343.

29. Li, Z., and Zhang, Y. (2006) Monodisperse silica-coated polyvinylpyrrolidone/$NaYF_4$ nanocrystals with multicolor upconversion fluorescence emission, *Angew. Chem. Int. Ed.*, **118**(46), 7896–7899.

30. Qian, H. S., *et al.* (2009) Mesoporous-silica-coated upconversion fluorescent nanoparticles for photodynamic therapy, *Small*, **5**(20), 2285–2290.

31. Lu, Q., *et al.* (2008) Silica-/titania-coated Y_2O_3:Tm^{3+}, Yb^{3+} nanoparticles with improvement in upconversion luminescence induced by different thickness shells, *J. Appl. Phys.*, **103**(12), 123533–123533-10.

32. Stouwdam, J. W., and van Veggel F. C. J. M. (2004) Improvement in the luminescence properties and processability of LaF_3/Ln and $LaPO_4$/Ln nanoparticles by surface modification, *Langmuir*, **20**(26), 11763–11771.

33. Xu, C. T., *et al.* (2008) Autofluorescence insensitive imaging using upconverting nanocrystals in scattering media, *Appl. Phys. Lett.*, **93**(17), 171103-3.

34. Shan, J., *et al.* (2009) NIR-to-visible upconversion nanoparticles for fluorescent labeling and targeted delivery of siRNA, *Nanotechnology*, **20**(15), 155101.

35. Chatterjee, D. K., Rufaihah, A. J., and Zhang, Y. (2008) Upconversion fluorescence imaging of cells and small animals using lanthanide doped nanocrystals, *Biomaterials*, **29**(7), 937–943.

36. Wang, M., *et al.* (2009) NIR-responsive silica-coated $NaYbF_4$: Er/Tm/Ho upconversion fluorescent nanoparticles with tunable emission colors and their applications in immunolabeling and fluorescent imaging of cancer cells, *J. Phys. Chem. C*, **113**(44), 19021–19027.

37. Zako, T., *et al.* (2009) Cyclic RGD peptide-labeled upconversion nanophosphors for tumor cell-targeted imaging, *Biochem. Biophys. Res. Commun.*, **381**(1), 54–58.

38. Abdul Jalil, R., and Zhang, Y. (2008) Biocompatibility of silica-coated $NaYF_4$ upconversion fluorescent nanocrystals, *Biomaterials*, **29**(30), 4122–4128.

39. Jiang, S., and Zhang, Y. (2010) Upconversion nanoparticle-based FRET system for study of siRNA in live cells, *Langmuir*, **26(9)**, 6689–6694.

40. Wu, S., *et al.* (2009) Non-blinking and photostable upconverted luminescence from single lanthanide-doped nanocrystals, *Proc. Natl. Acad. Sci. USA*, **106**(27), 10917–10921.

41. Wang, L., and Li, Y. (2007) Luminescent nanocrystals for nonenzymatic glucose concentration determination, *Chem. Eur. J.*, **13**(15), 4203–4207.

42. von Wallbrunn, A., *et al.* (2007) *In vivo* imaging of integrin αvβ3 expression using fluorescence-mediated tomography, *Eur. J. Nucl. Med. Mol. Imaging*, **34**(5), 745–754.

43. Lim, S. F., *et al.* (2005) *In vivo* and scanning electron microscopy imaging of upconverting nanophosphors in caenorhabditis elegans, *Nano Lett.*, **6**(2), 169–174.

44. Hischem, *et al.* (2009) *In-vivo* imaging of the uptake of upconversion nanoparticles by plant roots, *J. Biomed. Nanotechnol.*, **5**, 278–284.

45. Nyk, M., *et al.* (2008) High contrast *in vitro* and *in vivo* photoluminescence bioimaging using near infrared to near infrared up-conversion in Tm^{3+} and Yb^{3+} doped fluoride nanophosphors, *Nano Lett.*, **8(11)**, 3834–3838.

46. Hilderbrand, S. A., Salthouse, C., Mahmood, U., and Weissleder, R. (2009) Upconverting luminescent nanomaterials: application to *in vivo* bioimaging, *Chem. Commun.*, **28**, 4188–4190.

47. Vinegoni, C., *et al.* (2009) Transillumination fluorescence imaging in mice using biocompatible upconverting nanoparticles, *Opt. Lett.*, **34**(17), 2566–2568.

48. Li, Z., *et al.* (2009) Hybrid lanthanide nanoparticles with paramagnetic shell coated on upconversion fluorescent nanocrystals, *Langmuir*, **25**(20), 12015–12018.

49. Park, Y. I., *et al.* (2009) Nonblinking and nonbleaching upconverting nanoparticles as an optical imaging nanoprobe and T_1 magnetic resonance imaging contrast agent, *Adv. Mater.*, **21**(44), 4467–4471.

50. Ungun, B., *et al.* (2009) Nanofabricated upconversion nanoparticles for photodynamic therapy, *Opt. Express*,. **17**(1), 80–86.

51. Chatterjee, D. K., and Yong Z. (2008) Upconverting nanoparticles as nanotransducers for photodynamic therapy in cancer cells, *Nanomedicine*, **3**(1), 73–82.

52. Guo, H., *et al.* Singlet oxygen-induced apoptosis of cancer cells using upconversion fluorescent nanoparticles as a carrier of photosensitizer, *Nanomedicine*: *Nanotechnology, Biology and Medicine*.

Chapter 4

Rare Earth Nanomaterials in Magnetic Resonance Imaging

S. Roux,[a,b] R. Bazzi,[c,b] C. Rivière,[d] F. Lux,[a] P. Perriat,[e] and O. Tillement[a]

[a] *Laboratoire de Physico-Chimie des Matériaux Luminescents, UMR 5620 CNRS, Université Claude Bernard Lyon1, Villeurbanne, France*

[b] *Present address: Institut UTINAM, UMR 6213 CNRS, Université de Franche-Comté, Besançon, France*

[c] *Laboratoire de Physicochimie des Electrolytes, Colloïdes et Sciences Analytiques, UMR 7195 CNRS, Université Pierre et Marie Curie, Paris, France*

[d] *Laboratoire de Physique de la Matière Condensée et Nanostructures, UMR5586 CNRS, Université Claude Bernard Lyon1, Villeurbanne, France*

[e] *Groupe d'Etudes de Métallurgie Physique et de Physique des Matériaux, UMR 5510 CNRS — INSA de Lyon, 69621 Villeurbanne Cedex, France*

stephane.roux@univ-fcomte.fr

4.1 Introduction

The intense research activities devoted to nanoparticles during the last decade have opened the door to promising biological and medical applications [1–4]. Besides their reduced size, which makes them suitable for labeling biomolecules or for exploring the living machinery at the subcellular scale without functional alteration [5],

Rare Earth Nanotechnology
Edited by Timothy T.Y. Tan

ISBN 978-981-4316-30-9 (Hardcover), 978-981-4364-20-1 (eBook)
www.panstanford.com

the great potential of nanoparticles rests on the ability to gather in a same object several complementary properties. This was illustrated by the pioneering works of Weissleder's group that led to the development of triple label nanoparticles suited for the *in vivo* tracking of labeled cells [6]. They are composed of a superparamagnetic iron oxide core (5 nm) encapsulated in a dextran shell (thickness: 20 nm), which allows the derivatization of the particles both by fluorescent TAT peptides and by radioisotope ^{111}In chelates. The functionalization of these particles by TAT peptide facilitates their internalization in cells, which is successfully followed up by fluorescence imaging (FI), magnetic resonance imaging (MRI), and nuclear imaging. Such a combination of several detection techniques, that the use of multimodal particulate agents made possible, ensures a better reliability of the collected data. As a result, benefits can be expected for *in vivo* small-animal imaging, which is a prerequisite step before clinical application but also a very useful tool for biomedical investigation. Because of the great repercussions on clinical diagnosis and surgical protocol, the development of multimodal contrast agents for *in vivo* imaging is a rapid growing field [7]. Particularly, nanoparticles combining FI and MRI received much attention because they ally the high sensitivity of the fluorescence phenomenon to the high spatial resolution of MRI. Contrary to FI, whose *in vivo* application is seriously limited because of the weak penetration depth of light, MRI has become a prominent technique in diagnostic clinical medicine because of the possibility to get highly resolved three-dimensional images of living bodies [8–10]. Moreover, this technique contributes to improve the comfort of the patient because it is non-invasive, rapid, and avoids the use of radiochemicals [10].

Thanks to its highly resolved 3D noninvasive imaging capabilities, MRI appears as one of the most powerful clinical diagnostic techniques. Not only can MRI provide precise anatomical information, it can also provide functional measurements. MRI technique takes advantage of the magnetic resonance of water molecules that compose our body. The contrast obtained in magnetic resonance images is due to the differences in the density and relaxation time of water protons within the defined volume of investigation. While the density of water protons will depend on the level of hydration of the different structures of our body, the relaxation time will be correlated to the environment of water molecules and the degree of mobility of water molecules within this environment. MRI contrast is, thus, dependent on chemical, physical, and biological properties of the investigated tissue. Any changes in the tissue (such as protein density, presence of edema or tumors, etc.) will affect magnetic interactions between water protons, leading to changes in the MRI contrast properties of that tissue. The relaxation of water protons is also dependent on the strength of the static magnetic field and the pulse sequence of radio frequency waves. The development of various types of sequences has made it possible to obtain images mainly reflecting water protons density, transverse relaxation time (T_2 imaging), or longitudinal relaxation time (T_1 imaging). Signal tends to increase with decreasing relaxation time T_1 and decrease with decreasing relaxation time T_2. The interpretation of the resulting images, therefore, leads to the delineation and identification of most tissues.

To increase MRI sensitivity, contrast agents have been developed, which modify the relaxation time of water protons in their vicinity, hence modifying the resulting signal intensity detected in this region. Contrast agents are divided into two groups: negative and positive. Negative contrast agents are mainly superparamagnetic iron oxide nanoparticles, which induce a large shortening of the transverse relaxation time T_2 of surrounding water protons, resulting in a decrease in tissue signal intensity and, hence, darkening of the magnetic resonance images. Positive contrast agents are mainly paramagnetic chelates, which induce a large shortening of the longitudinal relaxation time T_1 of surrounding water protons, resulting in an increase in tissue signal intensity and, hence, brightening of the magnetic resonance images. Even if quantitative results can be obtained in theory for both negative and positive contrast agents, it requires development of important data treatment and physical interpretations for negative contrast agents, whereas it is much more straightforward for positive contrast agents.

Among positive contrast agents, gadolinium (III), a rare earth compound, is the most widely used paramagnetic element due to its seven unpaired electrons and relatively long electronic relaxation. Contrast properties are dependent on two key features: the water exchange rate between bulk water and water bound to the Gd ions and the rotational correlation time of the Gd-containing entities.

As free gadolinium is extremely toxic, gadolinium ions are caged within chelates structures such as DTPA (diethylenetriaminepentaacetic acid, Magnevist®), DOTA (1,4,7,10-tetraazacyclododecane-1,4,7,10-tetraacetic acid, Dotarem®), etc.

(Scheme 4.1) [8–10]. The caging limits the number of coordinated water molecules associated with each Gd ion and, hence, limits the modification of signal intensity that could be induced with such paramagnetic element. Another limitation is due to the small size (<1 nm) of clinically used Gd-complexes, which leads to short rotational correlation time, hence limiting the modification of proton relaxivity. Their small size also reduces the blood half-life of Gd-complexes, and blood-pool imaging is only accessible during the first minutes after intravenous injection of the contrast agent. Moreover, as it is quickly eliminated from the body, both passive (by taking advantage of Enhanced Permeability and Retention (EPR) effect in tumorous tissues) and active (by grafting relevant target-specific ligand on the surface of the contrast agents) targeting strategies are limited.

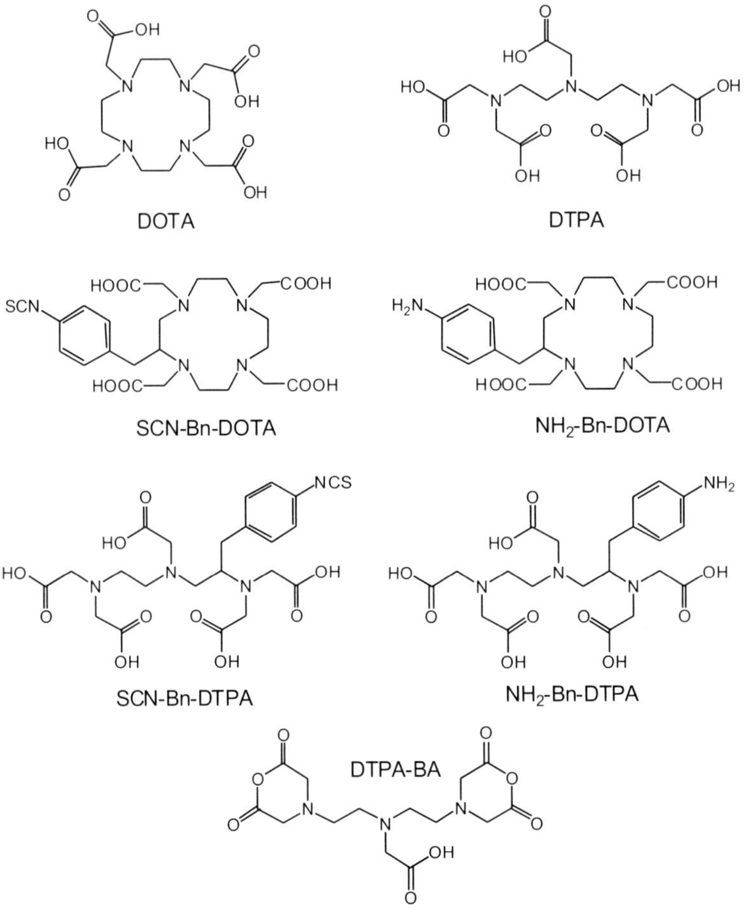

Scheme 4.1 The structures of the commercially available ligands.

In that context, many nanostructures have been developed in the past decades to (i) increase MRI sensitivity while keeping a positive contrast enhancement to get quantitative results more easily, and (ii) tune the biodistribution of the contrast agent for tissue-targeting purpose. Moreover, these nanostructures can often support additional attractive features to combine imaging and therapeutic functionalities and/or to induce a contrast in different complementary imaging techniques (optical, MRI, positron emission tomography (PET), ultrasound, etc.).

This chapter will focus specifically on the different nanostructures-based strategies employed to increase the sensitivity and specificity of the positive contrast agents.

To increase the sensitivity of rare earth based contrast agents, three parameters are usually considered: (i) increasing the number of Gd per contrast agent, (ii) increasing the rotational correlation time by increasing molecular weight and size, and (iii) increasing the number of coordinated water molecules.

The sensitivity of a given contrast agent is classically quantified by measurements of the longitudinal and transverse relaxivity values, r_1 and r_2, which refer to the amount of increase in $1/T_1$ and $1/T_2$, respectively, per millimolar of agent (often given as per mM of Gd). The ratio r_2/r_1 is used to classify the contrast agent as positive T_1 agents, which usually have r_2/r_1 ratios of 1–2, or as negative T_2 agents, which have r_2/r_1 ratios as high as 10 or more. For nanostructure elements, it is not only the relaxivity per Gd that defines the efficiency of the contrast agent but also the number of Gd per nanomaterial. It is then of interest to compute the relaxivity values per mM of nanomaterial, instead of per mM of Gd.

The first part of this chapter is dedicated to a review of different strategies to confer the ability to enhance the contrast of magnetic resonance images by grafting gadolinium chelates to various nanoparticles, while the second part is devoted to the development of rare earth elements containing crystalline nanoparticles (Scheme 4.2). It must be pointed out that the topic of this chapter is limited to "solid" nanoparticles, although a large range of vesicles, liposomes, and micelles have been successfully developed as contrast agents for MRI. Moreover, only a few examples of other rare earth elements (Dy, Ho), even if they exhibit an interesting potential for the design of particulate contrast agent for MRI, are given in this chapter since the data are rather rare as compared to particulate contrast agent based on gadolinium. This chapter will also introduce the use of such magnetic resonance contrast agents for both imaging and therapy in the final section.

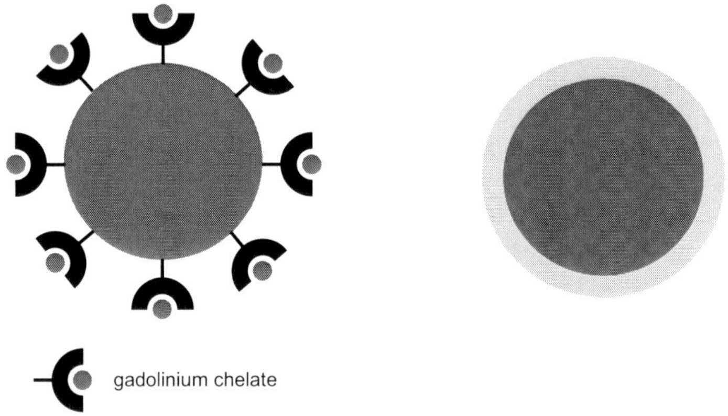

Scheme 4.2 Nanoparticle with a paramagnetic shell composed of gadolinium chelates (left) and nanoparticle with a paramagnetic crystalline core (right).

4.2 Multifunctional Nanostructures Functionalized by Gadolinium Chelates for Multi-Modal Imaging

4.2.1 Introduction

Surface derivatization of nanoparticle by gadolinium chelates constitutes an attractive route for the elaboration of highly sensitive MRI contrast agents. As already stated in the previous section, this strategy is expected to improve the enhancement of the positive contrast of magnetic resonance images in comparison to isolated gadolinium chelates because of the large amount of paramagnetic centers per probe and the reduction of the rotational motion. Since nanoparticles have the possibility to gather a large range of properties in a reduced volume, the immobilization of gadolinium chelates on nanoparticles increases the multifunctional character of the nanoparticles. Many studies have especially focused on the adjunction of MRI modality to fluorescent nanoparticles. The combination of FI and MRI in the same nanomaterial provides imaging probes characterized by a high sensitivity (due to the fluorescence properties) and high resolution (owing to the specifications of MRI). Fluorescence of these paramagnetic nanoparticles can arise from the encapsulation of organic dyes (extrinsic fluorescence) or from intrinsic properties of the nanoparticles. Among the latter, quantum dots are probably the most well known. Other combinations of imaging techniques can also be envisaged by functionalizing gold nanoparticles with gadolinium chelates since gold nanoparticles are able to induce a contrast enhancement in X-ray imaging. The association of several imaging modalities is very important because it ensures a better reliability of the observations and also allows monitoring the contrast agents at various scales (from the subcellular scale to the whole organism) by different techniques [6, 11–13]. Moreover, an increasing number of studies recently emphasized on the therapeutic activity of nanoparticles, which can be remotely controlled by an external physical stimulus (magnetic field, visible and near-infrared (NIR) light, X-ray photons, and neutron beam). In other words, harmless nanoparticles can be rendered cytotoxic after energy absorption [4, 14]. For an optimal effect, the follow up or visualization of these therapeutic agents is required from the administration to their elimination from the body. For instance, the activation of these nanoparticles must be performed only when they are accumulated, at least, preferentially in the pathological zone. The possibility to image the fate of the nanoparticles is consequently a crucial issue. Since most of these nanoparticles are not detected by MRI, surface functionalization by gadolinium chelates is, therefore, performed. However, efficient surface functionalization strategies depend on the chemical properties of the surface of nanoparticles. For this reason, the composition of the ligand and/or the surface must be tuned for ensuring an efficient anchoring of the gadolinium chelates.

The functionalization of nanoparticles aims to provide highly sensitive MRI contrast agents and/or the combination of several imaging techniques and/or the visualization of therapeutic agents. Most of the particles developed for eventual biomedical applications are essentially composed of silica (actually polysiloxane), quantum dots, and gold nanoparticles. The functionalization of nanoparticles by gadolinium chelate depends on the chemical composition of the surface, which is, in most cases, imposed by the nature of nanoparticles.

4.2.2 Fluorescent Silica Nanoparticles and Other Oxide Nanoparticles

Silica nanoparticles can be easily produced by applying either the protocol developed by Stöber or the microemulsion techniques. Stöber's protocol consists of alkaline hydrolysis–condensation of tetraethyl orthosilicate (TEOS) added in a large volume of ethanol [15]. The high dilution and the aqueous NH_4OH to TEOS ratio are key parameters for yielding monodisperse silica spheres in a large range of size (from 20 to 800 nm). The control of the size, when microemulsion techniques are applied, is ensured by the volume of micelles, which behave as nanoreactors [16]. Moreover, these synthesis protocols can be exploited for the encapsulation of metal oxide nanoparticles because of the great affinity between the surface of these particles and the polysiloxane shell precursors. Besides their simplicity, both methods are very attractive because multifunctional nanoparticles can be obtained by minor variations in the composition of the polysiloxane precursors and/or by post-functionalization. Indeed a large variety of polysiloxane precursors are available for envisaging an accurate control of the composition of the silica nanoparticles and for anticipating a functionalization before and/or after the hydrolysis-condensation. The production of multifunctional silica nanoparticles can be performed in several steps for the elaboration of multilayered nanostructures [17]. Each layer and the surface after derivatization contribute to the multifunctional character of the nanoparticles since they can each be designed for expressing a peculiar feature. For instance, aminopropyltriethoxysilane (APTES) was widely used in mixture with TEOS for introducing in silica nanoparticle, in each layer or on the surface amino groups, which can act as grafting site for the immobilization of organic dyes, hydrophilic molecules for improving the colloidal stability in aqueous biological media, biotargeting groups for increasing the accumulation of the nanoparticles in the zone of interest within the organism (Scheme 4.3). Van Blaaderen *et al.* demonstrated that the conjugation of fluorescein isothiocyanate to APTES before the hydrolysis–condensation allows the formation of fluorescent silica nanoparticles in which the organic dyes are distributed *either* in the whole nanoparticles or in the thickness of the polysiloxane shell, whereas the post-functionalization (i.e., after the hydrolysis–condensation) of the amino groups leads to the tethering of the organic dyes onto the nanoparticles [17]. Wiesner *et al.* developed the synthesis of highly fluorescent nanoparticles, which are composed of organic dyes (rhodamine B isothiocyanate (RBITC), cyanine 5 NHS ester (Cy5-NHS)) loaded core and a protective shell [18]. The fluorescent cores were obtained by adapting Stöber's protocol: the hydrolysis–condensation was performed on a mixture of TEOS and organic dyes conjugated to APTES. In the second step, the protective shell was obtained by hydrolysis-condensation of a mixture of TEOS and APTES whose amino group makes possible the covalent immobilization of neutral poly (ethylene glycol) chains. Such a post-functionalization favors the renal clearance of nanoparticles and significantly reduces the liver uptake after the *in vivo* injection of these highly fluorescent nanoprobes [19].

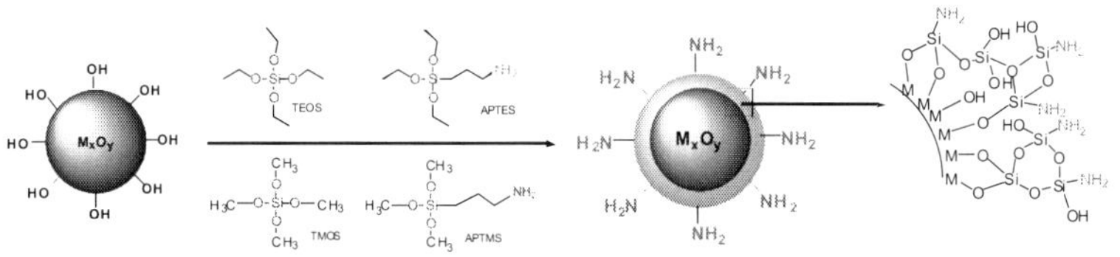

Scheme 4.3 Encapsulation of an oxide nanoparticle in an aminated polysiloxane shell obtained by hydrolysis-condensation of a mixture of TEOS + APTES or TMOS + APTMS.

The microemulsion technique can also be exploited for the synthesis of fluorescent silica nanoprobes. The research group of W. Tan determined the experimental conditions for the production of fluorescent nanoparticles in water in oil (W/O) microemulsion (reverse microemulsion). Water droplets (micelles) are dispersed in oil (cyclohexane) phase and stabilized by a surfactant layer (Triton X-100) for preventing the coalescence of droplets. The alkaline hydrolysis-condensation of TEOS and organic dyes conjugated APTES takes place inside the micelles. This group demonstrated the versatility of this strategy since various organic and inorganic dyes can be covalently or noncovalently encapsulated in the amorphous silica matrix [16, 20]. They succeeded to tune the color of the resulting nanoprobes and, therefore, to cover the visible spectral domain by embedding in the same nanoparticles several distinct organic dyes but with various ratios [21]. Moreover, these nanoparticles can be post-functionalized. In other words, the surface of these fluorescent nanoparticles can be derivatized by the covalent immobilization of oligonucleotide for lowering the detection threshold of hybridization detection [20] or by the condensation of negatively charged polysiloxane precursors for improving the colloidal stability [16].

This brief overview, which is not exhaustive, aims at drawing attention on the multiple possible ways for providing multifunctional silica nanoparticles. They were exploited for the synthesis of fluorescent and paramagnetic silica nanoparticles.

Fluorescent, radio-opaque, and paramagnetic silica nanoparticles were synthesized by applying reverse microemulsion (water-in-oil system) techniques. Soluble ruthenium (II) based fluorophores are encapsulated in a silica core, which is further embedded in paramagnetic and aminated shells inside water droplets of the reverse microemulsion. The paramagnetic character of these nanoparticles was conferred by the capture of gadolinium (III) ions since the fluorescent core is covalently functionalized by n-(trimethoxysilylpropyl)ethyldiamine (TSPETE) [22]. The water dispersity of these particles was ensured by the formation of an outer shell obtained by polymerization of TEOS, APTES, and 3-(trihydroxysilyl)propylmethylphosphonate (THPMP). The presence of negatively charged THPMP compensates the positive charge of APTES and generates colloidal stability in water because the functionalization by THPMP provides a high negative charge to the surface of the particles. Moreover, the presence of amine functions, resulting from APTES, can be used for bioconjugation of the particles. Due to their composition, these particles behave as *in vitro* contrast agents for FI (Ru(bpy) dyes), MRI (TSPETE-Gd), and computed tomography (CT) imaging (Ru and Gd elements). However, these particles showed a lower radio-opacity than Omnipaque (iodine-based organic compound) a widely used contrast agent for clinical CT because Ru and Gd exhibit an atomic number smaller than the one of iodine. On the other hand, these particles exert a higher contrast enhancement for MRI than in the case of the use of commercially available MRI contrast agent (Gadoteridol, Gd-HP-DO3A). Unfortunately, no data about the stability of the complex TSPETE-Gd were available although it is suspected to be relatively low because TSPETE molecules have only five reactive coordination sites. Despite these imperfections, the development of these particles is important because it opens the door to multimodal imaging applications based on gadolinium (III) chelates coated nanoparticles.

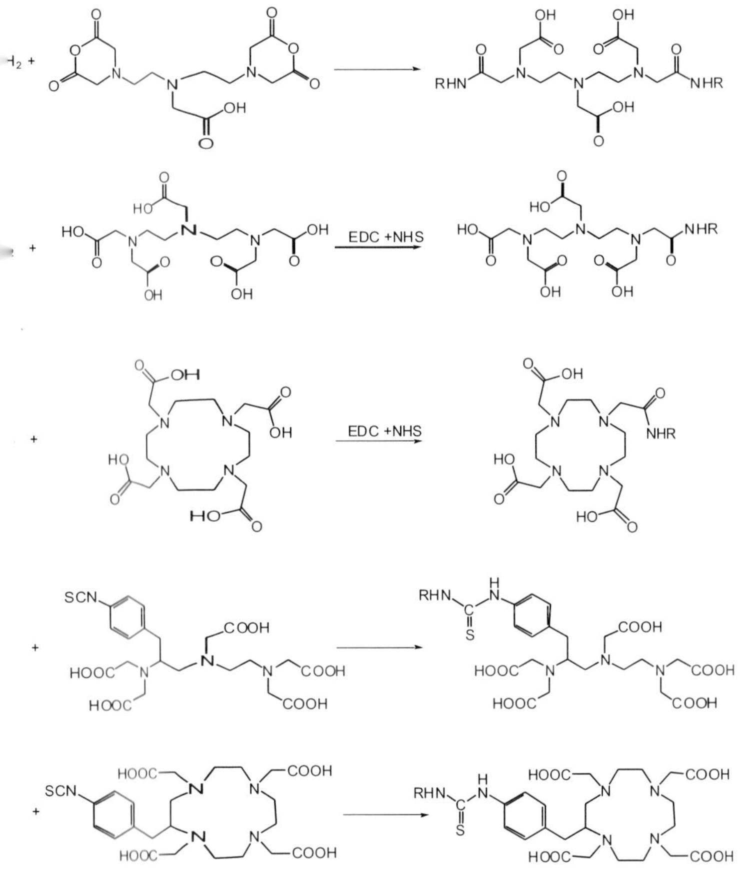

me 4.4 Reaction of DTPA, DOTA, and DTPA- and DOTA-analogues with amine groups.

ce and DTPA (Schemes 4.1, 4.4, 4.5). Despite a low amount of linium chelates immobilized on nanoparticles (four gadolinium per nanoparticle), a strong positive and negative contrast ncement is observed. This increase in relaxivity is consistent the increase in rotational correlation time, which results from the increase in molecular weight and maximum compactness e SiO_2 structure as compared to organic molecules. Indeed, igid nature of the inorganic nanoparticle tends to prevent the chelate from freely rotating or even folding back. Since the ensity of nanoparticles to be internalized in cells is higher in the case of the molecules, the immobilization of gadolinium tes onto nanoparticles is attractive because an efficient

Lin *et al.* performed the post-functionalization of fluorescent silica nanoparticles by two different gadolinium chelates, which carry one (Gd-Si-DTTA) or two (Gd-Si-DTPA) propyltrimethoxysilyl moieties [23]. After encapsulation of a luminescent $[Ru(bpy)_3]Cl_2$ core (bpy = 2,2'-bypyridine) in an amorphous silica matrix, the surface of the resulting nanoparticles was derivatized by condensation of silanol (Si-OH) groups present on the nanoparticles and the methoxysilyl groups of the gadolinium chelates. The authors noticed that the functionalization with Gd-Si-DTTA led to the formation of a monolayer of gadolinium chelates, whereas a thick paramagnetic multilayered shell was provided when Gd-Si-DTPA is used for the post-functionalization of the fluorescent nanoparticles. The final size of the nanoparticles is around 37 and 40 nm, respectively. The covalent immobilization of these gadolinium chelates confers to nanoparticles the ability to enhance the positive contrast of the MRI images since they exhibit higher longitudinal relaxivity values per mM Gd^{3+} ions than the one for unbound gadolinium complexes. Moreover each Gd-Si-DTTA chelate bound to the fluorescent silica nanoparticles exhibits higher relaxivities r_1 and r_2 than those of Gd-Si-DTPA grafted to similar nanoparticles (r_1 = 19.7 $mM^{-1}s^{-1}$ Gd^{3+} and r_2 = 60 $mM^{-1}s^{-1}$ Gd^{3+} vs r_1 = 7.8 $mM^{-1}s^{-1}$ Gd^{3+} and r_2 = 12.3 $mM^{-1}s^{-1}$ Gd^{3+}). This can be reasonably explained by a higher water exchange rate since gadolinium chelates Gd-Si-DTTA coordinate two water molecules whereas Gd-Si-DTPA chelates coordinate only one water molecule. In addition, another feature of nanoparticles coated by Gd-Si-DTTA plays in favor of a higher longitudinal relaxivity. Since the surface functionalization of the fluorescent silica nanoparticles by Gd-Si-DTTA yields a monolayered shell in contrast to the derivatization of the nanoparticles by Gd-Si-DTPA, gadolinium (III) ions are, therefore, more accessible. But the silica nanoparticles coated by Gd-Si-DTPA exhibit a higher longitudinal relaxivity r_1 when the comparison is based on the amount of particulate contrast agents since each particle coated by Gd-Si-DTPA contains more gadolinium (III) ions. It must be pointed out that these nanoparticles, despite having different coatings, have an impressive longitudinal relaxivity ($r_1 = 2.0 \times 10^5$ $mM^{-1}s^{-1}$ and $r_2 = 6.1 \times 10^5$ $mM^{-1}s^{-1}$ vs r_1 = 4.9×10^5 $mM^{-1}s^{-1}$ and $r_2 = 7.8 \times 10^5$ $mM^{-1}s^{-1}$ for Gd-Si-DTTA and Gd-Si-DTPA, respectively). The great potential of these luminescent and paramagnetic nanoparticles was confirmed by *in vitro* experiments carried out on monocyte cells with Gd-Si-DTTA-coated nanoparticles.

Their uptake by these cells was successfully and efficiently monitored by FI, flow cytometric experiments, and MRI (T_1 and T_2-weighted images). This study reveals that the internalization of nanoparticles in the monocyte cells does not alter the viability and that the fluorescent and paramagnetic nanoparticles allow multimodal *in vitro* imaging.

The authors explored another strategy for developing particulate contrast agents for imaging cells by FI and MRI. From fluorescent nanoparticles coated by Gd-Si-DTTA (see above), they applied a layer-by-layer (LbL) technique for immobilizing a higher amount of gadolinium chelates [24]. Since Gd-Si-DTTA coated nanoparticles are highly anionic, a layer of cationic Gd-DOTA oligomers was deposited onto the negatively charged surface via electrostatic interactions. As a result, the particles exhibit a positive charge that allows further deposition of anionic polystyrenesulfonate (PSS). The LbL assembly strategy, which consists in successive depositions of cationic Gd-DOTA oligomers and anionic PSS, yields luminescent and paramagnetic nanoparticles with alternate multilayer coatings of positively charged Gd-DOTA oligomers and negatively charged PSS. It was demonstrated that the size of the multilayered nanoparticles linearly increases with the number of deposition (from 37 nm (for a single layer) to 43 nm (for seven layers)). A similar evolution is observed for the relaxivities r_1 and r_2 of the particulate contrast agents because the alternate deposition of an additional cationic layer led to the increase in the amount of gadolinium (III) chelates for each nanoparticle. The per particle r_1 values increase from 1.94 $\times$ 10^5 $mM^{-1}s^{-1}$ for nanoparticles coated by a single Gd-Si-DTTA layer to 5.34 $\times$ 10^5 $mM^{-1}s^{-1}$ for Gd-Si-DTTA coated nanoparticles after the deposition of seven layers (four of cationic Gd-DOTA oligomers and three of anionic PSS), whereas the per particle r_2 values increase from 5.61 $\times$ 10^5 $mM^{-1}s^{-1}$ to 1.55 $\times$ 10^6 $mM^{-1}s^{-1}$. The LbL self-assembly thus offers an interesting strategy for increasing nanoparticle magnetic resonance relaxivities. It also affords the opportunity to design these particulate contrast agents for targeting cancerous cells, which is highly recommended for the early detection of tumor. Since LbL self-assembled nanoparticles are terminated with anionic PSS polymers, a biotargeting peptide sequence containing arginine-glycine-aspartate (RGD) and seven consecutive lysine (K) residues (K7RGD) can be immobilized onto the surface through electrostatic interaction between positively charged lysine residues of the

K7RGD sequence and negatively charged P
is well known for its strong interaction wit
upregulated in growing endothelial and a
Such a functionalization led to the efficient ta
cancer (HT-29) cells and calf pulmonary ar
cells as revealed by *in vitro* experiments m
Although these results are very promising, t
these contrast agents is not yet validated.

The silylated gadolinium chelates can als
for the functionalization of metal oxide n
are covered by –OH groups, which act as
al. reported the modification of boehmite (
with silylated gadolinium (III) chelates (de
These particles are, therefore, able to enha
of magnetic resonance images. As expect
are covalently bound to the nanoparticl
condensation between methoxy groups of s
groups of the surface, are able to capture
ions (Eu^{3+} and Tb^{3+}).

The post-functionalization of fluoresc
by silylated gadolinium (III) chelates
demonstrated by the studies of Lin's res
of the clinically approved gadolinium (III
polyaminocarboxylate motif, they can b
aminated nanoparticles through the for
(Schemes 4.4 and 4.5). Amine functio
generally introduced by the inorganic c
and aminated polysiloxane precursors (
or by the derivatization of oxide nanop
4.3). Delville *et al.* succeeded in conve
nanoparticles into fluorescent and para
[26]. This was performed after the coating
with a polysiloxane shell yielded by the
of APTMS. Such an encapsulation allow
(rhodamine isothiocyanate) and poly
(DTPA derivative) for the immobilizatio
onto the surface of the nanoparticles. DTP
surface either by the reaction between a
anhydride derivative of DTPA or by the p
by the carbodiimide EDC in presence o

labeling of microglial cells by paramagnetic agents is expected. The uptake of these particles by the cells was confirmed by the strong contrast enhancement and fluorescence. Microglial cells are well known to be chemoattracted and strongly accumulate particularly in brain pathological situations, including tumors. These cells can, therefore, behave as cargo for freighting a large amount of fluorescent and paramagnetic nanoparticles in the zone of interest. As a result, labeled cells can be followed up from the injection site to the target, which will be clearly delineated thanks to the presence of a large amount of contrast agents. This type of detection will open the route for more efficient contrast agents for magnetic resonance cell tracking.

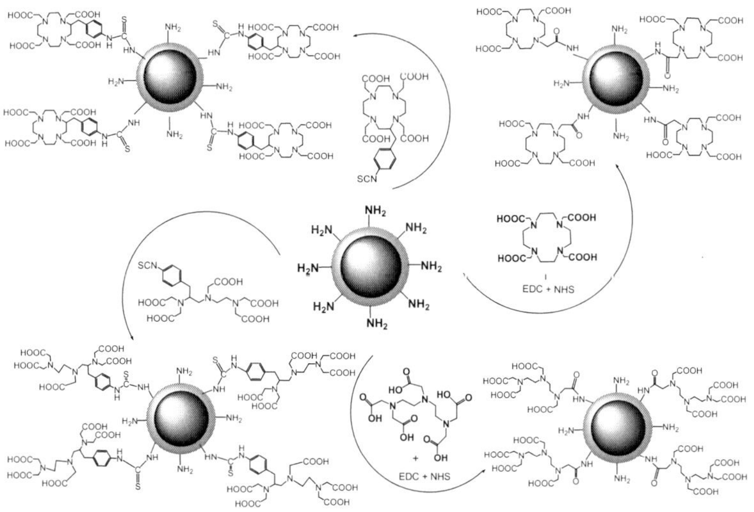

Scheme 4.5 Derivatization of aminated nanoparticles by DTPA, DOTA, and their analogues through the formation of thiourea (-$(NH)_2$-C=S) or amide (-NH-C(=O)-) linkages.

The aforementioned examples obviously show that the encapsulation of nanoparticles in a polysiloxane shell is an efficient route for rendering nanoparticles paramagnetic through the immobilization of gadolinium ions either by the use of silylated polyaminocarboxylate ligands or by the conjugation of polyaminocarboxylate ligands with amine functions that are present on the surface of the polysiloxane shell if aminated precursors are used. However, this method requires a strict control of the experimental conditions for avoiding the bulk polymerization of the polysiloxane precursors. The surface modification of transition metal oxide nanoparticles can also be performed by an alternative method, which relies on the grafting of gadolinium (III) chelates onto naked nanoparticles. These ligands are terminated by an anchoring function, which ensures their immobilization onto the nanoparticles since they are designed for establishing a strong interaction with the surface. The selective reactivity of the surface of TiO_2 nanoparticles to ortho-substituted enediols (such as dopamine) was exploited for grafting gadolinium (III) chelates (Scheme 4.6) [27]. Prior to the derivatization of TiO_2 nanoparticles, dopamine modified magnetic resonance contrast agents (DOPA-DO3A) was synthesized by coupling the commercially available succinimidyl ester activated 1,4,7,10-tetraazacyclododecane-1,4,7-tetraacetic acid to dopamine hydrochloride. The gadolinium (III) ions were captured by the resulting ligands. As expected, these particles are able to exert a contrast enhancement of magnetic resonance images since they exhibit a longitudinal relaxivity of 3.5 $mM^{-1}s^{-1}$ per gadolinium ion, which is similar to the value of clinically used small molecule contrast agents. Since this way of functionalization leads to the immobilization of a single monolayer of gadolinium(III) chelates, the longitudinal relaxivity per particle remains moderate (61 $mM^{-1}s^{-1}$) in comparison to the shell obtained from silylated ligands (up to 7.0×10^5 $mM^{-1}s^{-1}$, see above [23, 28]). In addition to dopamine-modified contrast agents, these nanoparticles can also be functionalized by DNA oligonucleotides (DNA-DOPA-DO3A-Gd) for targeted imaging and therapy. DNA-DOPA-DO3A-Gd nanoparticles were successfully applied for the labeling of PC3M cells, which were revealed by two-dimensional X-ray fluorescence maps and by T_1-weighted magnetic resonance images.

Dopamine and its derivatives were also used for the post-functionalization of superparamagnetic iron oxide (magnetite Fe_3O_4) nanoparticles because the spectroscopic study by Rajh [29] suggested that bidentate enediol ligands such as dopamine convert the undercoordinated Fe surface sites back to a bulk-like lattice structure with an octahedral geometry for oxygen-coordinated iron, which may result in tight binding of dopamine to iron oxide. Park *et al.* exploited the propensity of dopamine and its derivatives to be firmly bound to the iron oxide surface for grafting both hydrophilic polyethylene glycol chains and gadolinium chelates (Scheme 4.6) [30]. As a result, magnetite nanoparticles can be easily dispersed in deionized water and exhibit a behavior of dual contrast agents for T_1- and T_2-weighted MRI. Magnetite nanoparticles are indeed well known for their ability to induce negative contrast of magnetic resonance images (T_2 agents), whereas gadolinium chelates enhance the positive contrast of magnetic resonance images (T_1 agents).

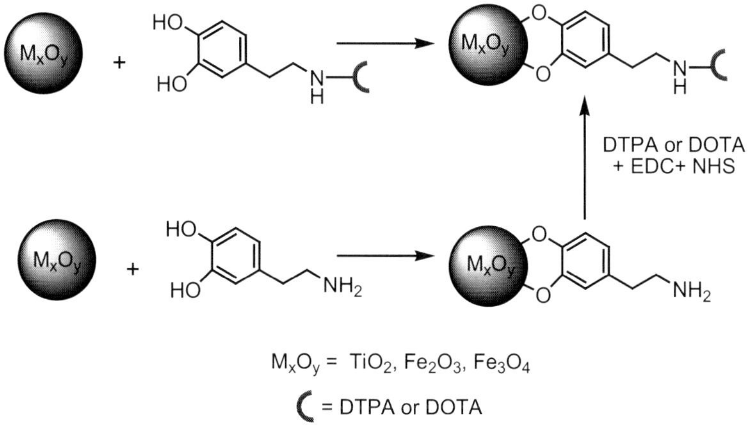

Scheme 4.6 Functionalization of titania and iron oxide nanoparticles by dopamine derivatives.

Since phosphonic acid groups have been found to be very effective in anchoring organic molecules to various metal oxide surfaces, a gadolinium (III) complex of polyaminocarboxylate ligands containing a phosphonate group in a pendant arm was used for the surface modification of TiO_2 nanoparticles [31].

4.2.3 Mesoporous Nanoparticles

Mesoporous nanoparticles are mainly based on silica nanoparticles. Such rigid structure has a large pore volume, with tunable pore size and surface functionalization. The biocompatibility of such system is now established [32, 33], and mesoporous silica nanoparticles are emerging as an ideal agent for biomedical application.

Mesoporous nanomaterials of tunable form (nanospheres [28, 34, 35], nanorods of various aspect ratios [36]), size (from 60 to 1100 nm) and pore size (from 2 to 10 nm) can be obtained [28]. Paramagnetic complexes were successfully incorporated in these structures for their use as MRI contrast agents.

The synthesis is based on a sol-gel procedure using surfactant such as CTAB (cetyltrimethylammonium bromide) as template. The first route consists in co-condensation under basic condition of tetraethyl orthosilicate (TEOS) with chelating agent (such as DTPA) conjugated with silane. The obtained solution of mesoporous nanostructure is then mixed with $GdCl_3$, which enables caging of Gd ions within the chelates [34, 36]. The second route consists of condensation of tetraethyl orthosilicate (TEOS) under basic condition. The obtained mesoporous structures are then coated with Gd-chelates conjugated with silane complex via a siloxane linkage [28].

As far as MRI contrast properties are concerned, the accessibility of paramagnetic entities to water molecules is one of the most important features to get high contrast agent relaxivities properties [23, 24]. The large surface-to-volume ratio of mesoporous nanomaterials appears to be an interesting property for the design of sensitive contrast agent. In fact, Gd-containing mesoporous nanomaterials present a remarkable relaxivity increase, compared to free Gd-complexes or Gd entrapped in plain silica nanoparticles. This enhancement in contrast properties is due to a concomitant effect of the porosity of the structure, which enables free movement of water molecules in and out of the structure, and of the interaction of Gd-complexes with the silica structure, which impedes the rotational movement of lanthanide complexes. This decrease in tumbling rate of the paramagnetic unit consequently increases the relaxivities of water molecules in the vicinity of the Gd-complex.

Five to ten fold increase in longitudinal and transverse relaxivities was reported for mesoporous nanorods of 100 nm with a 3/5 aspect ratio and a pore size of 2–3 nm [36], as well as on spherical mesoporous nanoparticles of 20–50 nm with pore size of 3.5 nm [37]. Lower enhancement on a per-Gd basis was reported for larger nanospheres (600–700 nm with pore size of 8 nm) [37]. The difference in relaxivities between various mesoporous nanostructures is attributed to the different accessibility of the water molecules to the Gd-complexes, localized inside or outside the silica pores [35]. It is hypothesized that only Gd-complexes grafted outside the nanopores contribute to the observed relaxivity, whereas the complexes grafted inside the nanopores do not modify the relaxation of water molecules. This may be due to a hindered or limited (slow) access of water molecules to the Gd-complexes inside the pores. As the localization of the Gd-complex on the inner or outer surfaces of the mesoporous silica particles strongly depends on the pore size, the corresponding relaxivity is similarly affected by the pore size.

Gd-load per nanomaterial is also key to increase the relaxivity in a per contrast-agent basis. However, the relation between Gd-load and relaxivity enhancement is not linear. Lin *et al.* reported a higher relaxivity for Gd-load ranging from 1.6% to 3.1% wt [38]. The r_1 value reaches its maximum for 1.6% Gd-load (23.6 mM^{-1} s^{-1} at 9.4 T and 23°C) and decreases abruptly as the Gd loading increases from 2.3% to 3.1%, falling to relaxivity values close to free Gd-DTPA at 6.8% (4.4 versus 3.9 $mM^{-1}s^{-1}$ for Gd-DTPA at 9.4 T and 23°C). At such high Gd loading, the dipole–dipole interaction among the paramagnetic ions is presumed to be more significant, hence shortening the electronic relaxation time.

The post-functionalization by silylated gadolinium chelates (Si-Gd-DTPA, see reference [23]) constitutes another attractive route for rendering mesoporous silica nanoparticles paramagnetic [28]. The immobilization of the gadolinium chelates is performed by the condensation of the trimethoxysilyl group of Si-Gd-DTPA and the –OH groups present on the pore walls and on the surface of the mesoporous nanoparticles. Such a post-functionalization confers to the mesoporous nanoparticles' very high longitudinal and transversal relaxivities r_1 and r_2, which amounts to 7.0×10^5 $mM^{-1}s^{-1}$ (3 T)/2.48×10^5 $mM^{-1}s^{-1}$ (9.4 T) and 1.6×10^6 mM^{-1} s^{-1} (3 T)/2.7×10^6 $mM^{-1}s^{-1}$ (9.4 T) on a per millimolar particle basis, respectively. These relaxivity values are much larger than the solid silica nanoparticles coated with multilayers of the Gd-DTPA derivative (see reference [23]). The enhanced magnetic resonance relaxivity was attributed to the ready access of water molecules through the nanochannels of mesoporous particles derivatized by gadolinium chelates Si-Gd-DTTA. When the functionalization is performed by a mixture of Si-Gd-DTTA and of rhodamine B conjugated to aminopropyltriethoxysilane (4 mol %, relative to Gd-Si-DTTA), the resulting mesoporous nanoparticles can behave as contrast agents for fluorescence and magnetic resonance imaging. The uptake of these fluorescent and paramagnetic mesoporous nanoparticles by monocyte cells was revealed both by the fluorescence and the contrast enhancement magnetic resonance images. Moreover, they appeared well suited for *in vivo* MRI after intravenous injection in the tail vein since aorta and liver are clearly visible on T_1- and

T_2-weighted images, respectively. These nanoparticles are indeed able to induce a positive and a negative contrast. If hepatic uptake renders liver imaging possible, thanks to the phagocytosis of the nanoparticles by the resident liver macrophage cells (Kupfer cells), it, however, leads to a quick clearance from the blood and limits the use of these nanoparticles as *in vivo* contrast agents.

Another interesting approach is to combine the incorporation of Gd with other lanthanide such as Eu and Tb since they are coordinated by the same ligands [38]. Eu^{3+} and Tb^{3+} possess excellent luminescent properties in the visible range. The post-functionalization of a fluorescent silica core, which contains Ru(bpy) as fluorophore with silylated polyaminocarboxylate ligands, allows the immobilization of both Tb^{3+} and Gd^{3+} ions. The presence of Tb^{3+} confers to the particles emission properties, which are complementary to those of Ru^{2+} entrapped inside the core, whereas gadolinium ions are expected to render the nanoparticles paramagnetic [39]. This approach enables the design of multimodal contrast agent that combine high resolution and sensitivity of optical set-up, with 3D localization provided by MRI.

The use of mesoporous nanoparticles as cell markers has also been studied. These structures are readily internalized by many type of cells, without the need of any transfection agent [28, 33, 36, 40]. Hsiao *et al.* have recently reported the use of hybrid mesoporous nanoparticles with Gd for stem cell tracking and were able to track as low as 4000 stem cells for 14 days using a 1.5T clinical MRI [34]. However, precise study on the effect of internalization on the contrast properties has to be performed for long-term cell trafficking visualization [41, 42].

4.2.4 Paramagnetic Quantum Dots

The functionalization of quantum dots (QD) by gadolinium chelates represents another route for the elaboration of fluorescent positive contrast agents combining FI and MRI.

QD refers to a crystallite cluster of II–VI semiconductors (CdS, CdSe, ZnS, CdSe@ZnS, etc.) exhibiting unique optical properties that arise from quantum confinement effects characteristic of these small dimensions (1–8 nm) [43–45]. Their attractive features (high photostability, high quantum yield, controllable and narrow emission band, broad excitation band) should contribute to a breakthrough in FI even if their routine application is still controversial (owing to the toxicity of Cd^{2+} ions) [46, 47]. Their relatively narrow emission bands, whose position is controlled either by the particle size or, for a given size, by tuning the composition of alloyed semiconducting QD (CdSeTe), together with their broad absorption spectra, offer a large palette of distinct color enabling multilabeling [48]. Since several synthetic routes have been developed for the preparation of water-soluble QD with narrow size distributions and high quantum yields, the literature is overflowing with examples of biological materials labeled by QD [49, 50]. Indeed the most common synthetic routes of QD lead to hydrophobic fluorescent nanoparticles since they are coated by trioctylphosphine/trioctylphosphine oxide (TOP/TOPO). The biological application of QD requires, therefore, their functionalization by capping hydrophilic molecules. Besides water dispersity, the functionalization plays two supplementary important roles: capping ligands can (i) serve as anchoring sites for grafting molecules and (ii) ensure, by passivating the QD surface, their protection and consequently the preservation of optical properties. A representative list of caps and the QD dispersal strategies they use is provided in a review written by the research group of Mattoussi [5]. These strategies can be grouped into three major routes (Scheme 4.7). The first one consists in the exchange of hydrophobic monolayer of TOP/TOPO by hydrophilic thiol or phosphine terminated ligands (mercaptocarboxylic acid, alkylthiol-terminated DNA, thioalkylated oligo-ethyleneglycols). This functionalization is driven by mass action and is possible because sulphur or phosphorus atom can establish strong dative bond with the chalcogenide species present at the surface of QD. However, it must be pointed out that the colloidal and chemical stability is considerably improved when hydrophilic ligands used for replacing TOP/TOPO carry at least two thiol or phosphine functions (dihydrolipoic acid derivatives, oligomeric phosphines). The second strategy described in the literature is based on the encapsulation of QD in a silica shell. As aforementioned (see above, Section 4.2.2), silica shells afford multiple ways for further surface modification. The first step of the formation of silica shell lies in the replacement of TOP/TOPO by mercaptopropyltrimethoxysilane (MPTMS) whose thiol end allows the immobilization onto the QD. The growth of the silica shell is, therefore, generated by hydrolysis-condensation of polysiloxane precursors mixture (TEOS, APTES, MPTMS, etc.) from the alkoxysilyl groups of MPTMS bound to the

QD surface. The third method preserves the native TOP/TOPO on the QD and uses variants of amphiphilic "diblock" and "triblock" copolymers and phospholipids to tightly interleave/interdigitate the alkylphosphine ligands through hydrophobic attraction, whereas the hydrophilic outer block permits aqueous dispersion and further derivitization.

These three strategies (Scheme 4.7) were applied for derivatizing fluorescent QD with gadolinium chelates. To add the MRI contrast ability, QD can be functionalized with Gd^{3+}-DOTA (1,4,7,10-tetraazacyclododecane-1,4,7,10-tetraacetic acid) complexes as shown by the study of Jin *et al.* [51]. For achieving this, TOP/TOPO layer was first replaced by glutathione, which is composed of a thiol function for the immobilization of the ligand onto the NIR-emitting QD, an amine group for further functionalization and of two carboxylic moieties for water compatibility and also for further functionalization. The addition of DOTA NHS ester to glutathione-coated QD led to the covalent grafting of DOTA through the formation of amide bond (Schemes 4.4 and 4.5). In a last functionalization step, QD are made paramagnetic by the capture of gadolinium (III) ions from DOTA (Gd-DOTA-QD). The functionalization induces an increase in the hydrodynamic diameter from 7 to 10 nm. This strategy seems to be efficient since Gd-DOTA-QD exhibit a relatively high longitudinal relaxivity r_1 (365 $mM^{-1}s^{-1}$). Moreover, these paramagnetic QD have a great potential for *in vivo* imaging since the implantation of a tube filled with Gd-DOTA-QD into the abdomen of a mouse can be detected by NIR FI and by MRI. Unfortunately, no data were given about the intravenous injection of these nanopaticles and their biodistribution, which is a key parameter for the biomedical applications of these QD.

Gerion *et al.* demonstrated that QD can be functionalized by gadolinium(III) chelates after their encapsulation in a silica shell [52]. The shell was obtained in two steps. After a priming step replacing trioctylphosphine oxide (TOPO) surfactants on the QD surface with mercaptopropyltrimethoxysilane (MPTMS), polymerization of siloxane was performed in methanol under slightly basic conditions. In the second step, addition of fresh MPTMS and poly(ethylene) glycol(PEG)-propyltrimethoxysilane permitted the introduction of functional (SH) and stabilizing PEG groups on the surface of the SiO_2 shell. As reported in the section devoted to fluorescent silica nanoparticles (see Section 4.2.2), ligands used for the immobilization

Lin *et al.* performed the post-functionalization of fluorescent silica nanoparticles by two different gadolinium chelates, which carry one (Gd-Si-DTTA) or two (Gd-Si-DTPA) propyltrimethoxysilyl moieties [23]. After encapsulation of a luminescent [Ru(bpy)$_3$]Cl$_2$ core (bpy = 2,2'-bypyridine) in an amorphous silica matrix, the surface of the resulting nanoparticles was derivatized by condensation of silanol (Si-OH) groups present on the nanoparticles and the methoxysilyl groups of the gadolinium chelates. The authors noticed that the functionalization with Gd-Si-DTTA led to the formation of a monolayer of gadolinium chelates, whereas a thick paramagnetic multilayered shell was provided when Gd-Si-DTPA is used for the post-functionalization of the fluorescent nanoparticles. The final size of the nanoparticles is around 37 and 40 nm, respectively. The covalent immobilization of these gadolinium chelates confers to nanoparticles the ability to enhance the positive contrast of the MRI images since they exhibit higher longitudinal relaxivity values per mM Gd^{3+} ions than the one for unbound gadolinium complexes. Moreover each Gd-Si-DTTA chelate bound to the fluorescent silica nanoparticles exhibits higher relaxivities r_1 and r_2 than those of Gd-Si-DTPA grafted to similar nanoparticles (r_1 = 19.7 mM^{-1}s^{-1} Gd^{3+} and r_2 = 60 mM^{-1}s^{-1} Gd^{3+} vs r_1 = 7.8 mM^{-1}s^{-1} Gd^{3+} and r_2 = 12.3 mM^{-1}s^{-1} Gd^{3+}). This can be reasonably explained by a higher water exchange rate since gadolinium chelates Gd-Si-DTTA coordinate two water molecules whereas Gd-Si-DTPA chelates coordinate only one water molecule. In addition, another feature of nanoparticles coated by Gd-Si-DTTA plays in favor of a higher longitudinal relaxivity. Since the surface functionalization of the fluorescent silica nanoparticles by Gd-Si-DTTA yields a monolayered shell in contrast to the derivatization of the nanoparticles by Gd-Si-DTPA, gadolinium (III) ions are, therefore, more accessible. But the silica nanoparticles coated by Gd-Si-DTPA exhibit a higher longitudinal relaxivity r_1 when the comparison is based on the amount of particulate contrast agents since each particle coated by Gd-Si-DTPA contains more gadolinium (III) ions. It must be pointed out that these nanoparticles, despite having different coatings, have an impressive longitudinal relaxivity ($r_1 = 2.0 \times 10^5$ mM^{-1}s^{-1} and $r_2 = 6.1 \times 10^5$ mM^{-1}s^{-1} vs r_1 = 4.9×10^5 mM^{-1}s^{-1} and $r_2 = 7.8 \times 10^5$ mM^{-1}s^{-1} for Gd-Si-DTTA and Gd-Si-DTPA, respectively). The great potential of these luminescent and paramagnetic nanoparticles was confirmed by *in vitro* experiments carried out on monocyte cells with Gd-Si-DTTA-coated nanoparticles.

Their uptake by these cells was successfully and efficiently monitored by FI, flow cytometric experiments, and MRI (T_1 and T_2-weighted images). This study reveals that the internalization of nanoparticles in the monocyte cells does not alter the viability and that the fluorescent and paramagnetic nanoparticles allow multimodal *in vitro* imaging.

The authors explored another strategy for developing particulate contrast agents for imaging cells by FI and MRI. From fluorescent nanoparticles coated by Gd-Si-DTTA (see above), they applied a layer-by-layer (LbL) technique for immobilizing a higher amount of gadolinium chelates [24]. Since Gd-Si-DTTA coated nanoparticles are highly anionic, a layer of cationic Gd-DOTA oligomers was deposited onto the negatively charged surface via electrostatic interactions. As a result, the particles exhibit a positive charge that allows further deposition of anionic polystyrenesulfonate (PSS). The LbL assembly strategy, which consists in successive depositions of cationic Gd-DOTA oligomers and anionic PSS, yields luminescent and paramagnetic nanoparticles with alternate multilayer coatings of positively charged Gd-DOTA oligomers and negatively charged PSS. It was demonstrated that the size of the multilayered nanoparticles linearly increases with the number of deposition (from 37 nm (for a single layer) to 43 nm (for seven layers)). A similar evolution is observed for the relaxivities r_1 and r_2 of the particulate contrast agents because the alternate deposition of an additional cationic layer led to the increase in the amount of gadolinium (III) chelates for each nanoparticle. The per particle r_1 values increase from 1.94×10^5 $mM^{-1}s^{-1}$ for nanoparticles coated by a single Gd-Si-DTTA layer to 5.34×10^5 $mM^{-1}s^{-1}$ for Gd-Si-DTTA coated nanoparticles after the deposition of seven layers (four of cationic Gd-DOTA oligomers and three of anionic PSS), whereas the per particle r_2 values increase from 5.61×10^5 $mM^{-1}s^{-1}$ to 1.55×10^6 $mM^{-1}s^{-1}$. The LbL self-assembly thus offers an interesting strategy for increasing nanoparticle magnetic resonance relaxivities. It also affords the opportunity to design these particulate contrast agents for targeting cancerous cells, which is highly recommended for the early detection of tumor. Since LbL self-assembled nanoparticles are terminated with anionic PSS polymers, a biotargeting peptide sequence containing arginine-glycine-aspartate (RGD) and seven consecutive lysine (K) residues (K7RGD) can be immobilized onto the surface through electrostatic interaction between positively charged lysine residues of the

K7RGD sequence and negatively charged PSS layer. RGD sequence is well known for its strong interaction with $\alpha_v\beta_3$ integrin that are upregulated in growing endothelial and angiogenic cancer cells. Such a functionalization led to the efficient targeting of human colon cancer (HT-29) cells and calf pulmonary artery endothelial (CPAE) cells as revealed by *in vitro* experiments monitored by FI and MRI. Although these results are very promising, the *in vivo* application of these contrast agents is not yet validated.

The silylated gadolinium chelates can also be successfully applied for the functionalization of metal oxide nanoparticles since they are covered by –OH groups, which act as anchoring sites. Frias *et al.* reported the modification of boehmite (AlO(OH)) nanoparticles with silylated gadolinium (III) chelates (derivatives of DO3A) [25]. These particles are, therefore, able to enhance the positive contrast of magnetic resonance images. As expected, these ligands, which are covalently bound to the nanoparticle surface through the condensation between methoxy groups of silylated ligands and –OH groups of the surface, are able to capture luminescent lanthanide ions (Eu^{3+} and Tb^{3+}).

The post-functionalization of fluorescent silica nanoparticles by silylated gadolinium (III) chelates is very efficient, as demonstrated by the studies of Lin's research group. Since most of the clinically approved gadolinium (III) chelates are based on polyaminocarboxylate motif, they can be easily immobilized on aminated nanoparticles through the formation of amide linkage (Schemes 4.4 and 4.5). Amine functions in nanoparticles are generally introduced by the inorganic copolymerization of TEOS and aminated polysiloxane precursors (APP like ATPES, APTMS) or by the derivatization of oxide nanoparticles by APP (Scheme 4.3). Delville *et al.* succeeded in converting alumina and silica nanoparticles into fluorescent and paramagnetic contrast agents [26]. This was performed after the coating of the bare nanoparticles with a polysiloxane shell yielded by the hydrolysis-condensation of APTMS. Such an encapsulation allows tethering organic dyes (rhodamine isothiocyanate) and polyaminocarboxylate ligand (DTPA derivative) for the immobilization of gadolinium (III) ions onto the surface of the nanoparticles. DTPA ligand was grafted to the surface either by the reaction between aminated surface and bis-anhydride derivative of DTPA or by the peptide reaction promoted by the carbodiimide EDC in presence of NHS between aminated

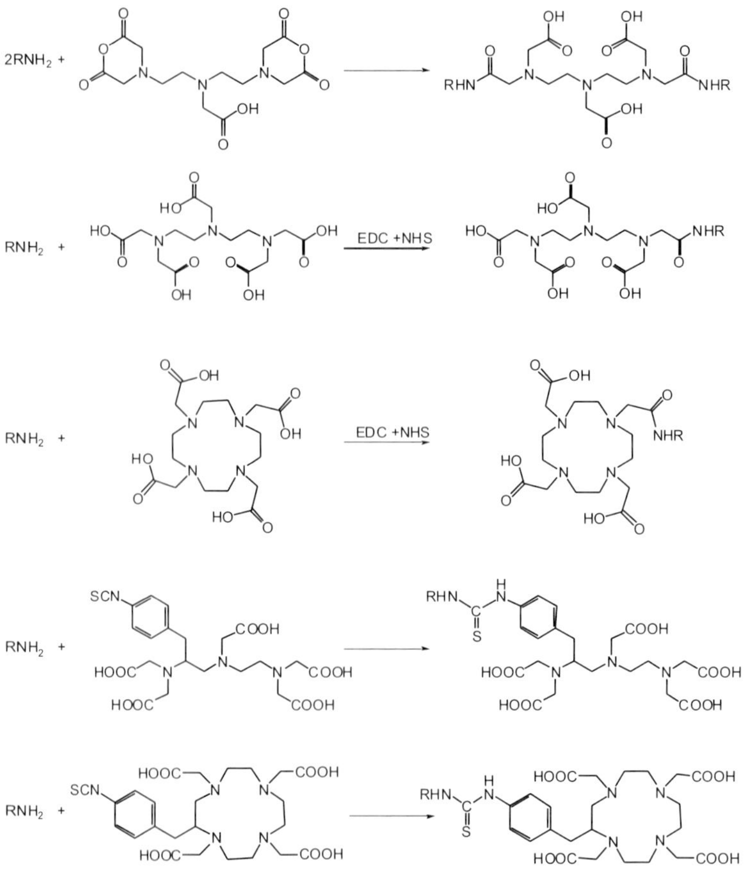

Scheme 4.4 Reaction of DTPA, DOTA, and DTPA- and DOTA-analogues with amine groups.

surface and DTPA (Schemes 4.1, 4.4, 4.5). Despite a low amount of gadolinium chelates immobilized on nanoparticles (four gadolinium ions per nanoparticle), a strong positive and negative contrast enhancement is observed. This increase in relaxivity is consistent with the increase in rotational correlation time, which results from both the increase in molecular weight and maximum compactness of the SiO_2 structure as compared to organic molecules. Indeed, the rigid nature of the inorganic nanoparticle tends to prevent the Gd^{3+} chelate from freely rotating or even folding back. Since the propensity of nanoparticles to be internalized in cells is higher than in the case of the molecules, the immobilization of gadolinium chelates onto nanoparticles is attractive because an efficient labeling of microglial cells by paramagnetic agents is expected. The uptake of these particles by the cells was confirmed by the strong contrast enhancement and fluorescence. Microglial cells are well known to be chemoattracted and strongly accumulate particularly in brain pathological situations, including tumors. These cells can, therefore, behave as cargo for freighting a large amount of fluorescent and paramagnetic nanoparticles in the zone of interest. As a result, labeled cells can be followed up from the injection site to the target, which will be clearly delineated thanks to the presence of a large amount of contrast agents. This type of detection will open the route for more efficient contrast agents for magnetic resonance cell tracking.

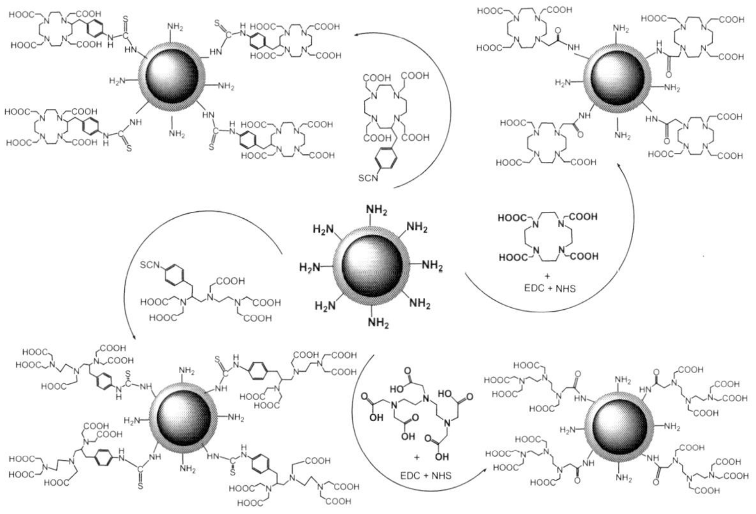

Scheme 4.5 Derivatization of aminated nanoparticles by DTPA, DOTA, and their analogues through the formation of thiourea (-$(NH)_2$-C=S) or amide (-NH-C(=O)-) linkages.

The aforementioned examples obviously show that the encapsulation of nanoparticles in a polysiloxane shell is an efficient route for rendering nanoparticles paramagnetic through the immobilization of gadolinium ions either by the use of silylated polyaminocarboxylate ligands or by the conjugation of polyaminocarboxylate ligands with amine functions that are present on the surface of the polysiloxane shell if aminated precursors are used. However, this method requires a strict control of the experimental conditions for avoiding the bulk polymerization of the polysiloxane precursors. The surface modification of transition metal oxide nanoparticles can also be performed by an alternative method, which relies on the grafting of gadolinium (III) chelates onto naked nanoparticles. These ligands are terminated by an anchoring function, which ensures their immobilization onto the nanoparticles since they are designed for establishing a strong interaction with the surface. The selective reactivity of the surface of TiO_2 nanoparticles to ortho-substituted enediols (such as dopamine) was exploited for grafting gadolinium (III) chelates (Scheme 4.6) [27]. Prior to the derivatization of TiO_2 nanoparticles, dopamine modified magnetic resonance contrast agents (DOPA-DO3A) was synthesized by coupling the commercially available succinimidyl ester activated 1,4,7,10-tetraazacyclododecane-1,4,7-tetraacetic acid to dopamine hydrochloride. The gadolinium (III) ions were captured by the resulting ligands. As expected, these particles are able to exert a contrast enhancement of magnetic resonance images since they exhibit a longitudinal relaxivity of 3.5 $mM^{-1}s^{-1}$ per gadolinium ion, which is similar to the value of clinically used small molecule contrast agents. Since this way of functionalization leads to the immobilization of a single monolayer of gadolinium(III) chelates, the longitudinal relaxivity per particle remains moderate (61 $mM^{-1}s^{-1}$) in comparison to the shell obtained from silylated ligands (up to 7.0×10^5 $mM^{-1}s^{-1}$, see above [23, 28]). In addition to dopamine-modified contrast agents, these nanoparticles can also be functionalized by DNA oligonucleotides (DNA-DOPA-DO3A-Gd) for targeted imaging and therapy. DNA-DOPA-DO3A-Gd nanoparticles were successfully applied for the labeling of PC3M cells, which were revealed by two-dimensional X-ray fluorescence maps and by T_1-weighted magnetic resonance images.

Dopamine and its derivatives were also used for the post-functionalization of superparamagnetic iron oxide (magnetite Fe_3O_4) nanoparticles because the spectroscopic study by Rajh [29] suggested that bidentate enediol ligands such as dopamine convert the undercoordinated Fe surface sites back to a bulk-like lattice structure with an octahedral geometry for oxygen-coordinated iron, which may result in tight binding of dopamine to iron oxide. Park *et al.* exploited the propensity of dopamine and its derivatives to be firmly bound to the iron oxide surface for grafting both hydrophilic polyethylene glycol chains and gadolinium chelates (Scheme 4.6) [30]. As a result, magnetite nanoparticles can be easily dispersed in deionized water and exhibit a behavior of dual contrast agents for T_1- and T_2-weighted MRI. Magnetite nanoparticles are indeed well known for their ability to induce negative contrast of magnetic resonance images (T_2 agents), whereas gadolinium chelates enhance the positive contrast of magnetic resonance images (T_1 agents).

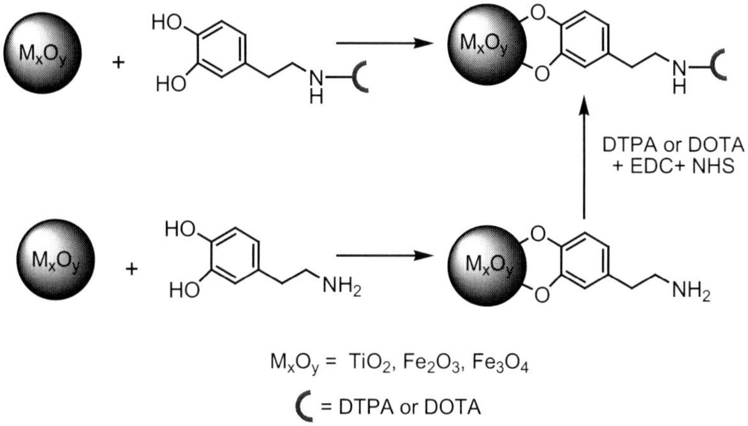

Scheme 4.6 Functionalization of titania and iron oxide nanoparticles by dopamine derivatives.

Since phosphonic acid groups have been found to be very effective in anchoring organic molecules to various metal oxide surfaces, a gadolinium (III) complex of polyaminocarboxylate ligands containing a phosphonate group in a pendant arm was used for the surface modification of TiO_2 nanoparticles [31].

4.2.3 Mesoporous Nanoparticles

Mesoporous nanoparticles are mainly based on silica nanoparticles. Such rigid structure has a large pore volume, with tunable pore size and surface functionalization. The biocompatibility of such system is now established [32, 33], and mesoporous silica nanoparticles are emerging as an ideal agent for biomedical application.

Mesoporous nanomaterials of tunable form (nanospheres [28, 34, 35], nanorods of various aspect ratios [36]), size (from 60 to 1100 nm) and pore size (from 2 to 10 nm) can be obtained [28]. Paramagnetic complexes were successfully incorporated in these structures for their use as MRI contrast agents.

The synthesis is based on a sol-gel procedure using surfactant such as CTAB (cetyltrimethylammonium bromide) as template. The first route consists in co-condensation under basic condition of tetraethyl orthosilicate (TEOS) with chelating agent (such as DTPA) conjugated with silane. The obtained solution of mesoporous nanostructure is then mixed with $GdCl_3$, which enables caging of Gd ions within the chelates [34, 36]. The second route consists of condensation of tetraethyl orthosilicate (TEOS) under basic condition. The obtained mesoporous structures are then coated with Gd-chelates conjugated with silane complex via a siloxane linkage [28].

As far as MRI contrast properties are concerned, the accessibility of paramagnetic entities to water molecules is one of the most important features to get high contrast agent relaxivities properties [23, 24]. The large surface-to-volume ratio of mesoporous nanomaterials appears to be an interesting property for the design of sensitive contrast agent. In fact, Gd-containing mesoporous nanomaterials present a remarkable relaxivity increase, compared to free Gd-complexes or Gd entrapped in plain silica nanoparticles. This enhancement in contrast properties is due to a concomitant effect of the porosity of the structure, which enables free movement of water molecules in and out of the structure, and of the interaction of Gd-complexes with the silica structure, which impedes the rotational movement of lanthanide complexes. This decrease in tumbling rate of the paramagnetic unit consequently increases the relaxivities of water molecules in the vicinity of the Gd-complex.

Five to ten fold increase in longitudinal and transverse relaxivities was reported for mesoporous nanorods of 100 nm with a 3/5 aspect ratio and a pore size of 2–3 nm [36], as well as on spherical mesoporous nanoparticles of 20–50 nm with pore size of 3.5 nm [37]. Lower enhancement on a per-Gd basis was reported for larger nanospheres (600–700 nm with pore size of 8 nm) [37]. The difference in relaxivities between various mesoporous nanostructures is attributed to the different accessibility of the water molecules to the Gd-complexes, localized inside or outside the silica pores [35]. It is hypothesized that only Gd-complexes grafted outside the nanopores contribute to the observed relaxivity, whereas the complexes grafted inside the nanopores do not modify the relaxation of water molecules. This may be due to a hindered or limited (slow) access of water molecules to the Gd-complexes inside the pores. As the localization of the Gd-complex on the inner or outer surfaces of the mesoporous silica particles strongly depends on the pore size, the corresponding relaxivity is similarly affected by the pore size.

Gd-load per nanomaterial is also key to increase the relaxivity in a per contrast-agent basis. However, the relation between Gd-load and relaxivity enhancement is not linear. Lin *et al.* reported a higher relaxivity for Gd-load ranging from 1.6% to 3.1% wt [38]. The r_1 value reaches its maximum for 1.6% Gd-load (23.6 $mM^{-1}\ s^{-1}$ at 9.4 T and 23°C) and decreases abruptly as the Gd loading increases from 2.3% to 3.1%, falling to relaxivity values close to free Gd-DTPA at 6.8% (4.4 versus 3.9 $mM^{-1}s^{-1}$ for Gd-DTPA at 9.4 T and 23°C). At such high Gd loading, the dipole–dipole interaction among the paramagnetic ions is presumed to be more significant, hence shortening the electronic relaxation time.

The post-functionalization by silylated gadolinium chelates (Si-Gd-DTPA, see reference [23]) constitutes another attractive route for rendering mesoporous silica nanoparticles paramagnetic [28]. The immobilization of the gadolinium chelates is performed by the condensation of the trimethoxysilyl group of Si-Gd-DTPA and the –OH groups present on the pore walls and on the surface of the mesoporous nanoparticles. Such a post-functionalization confers to the mesoporous nanoparticles' very high longitudinal and transversal relaxivities r_1 and r_2, which amounts to 7.0×10^5 $mM^{-1}s^{-1}$ (3 T)/2.48×10^5 $mM^{-1}s^{-1}$ (9.4 T) and 1.6×10^6 $mM^{-1}\ s^{-1}$ (3 T)/2.7×10^6 $mM^{-1}s^{-1}$ (9.4 T) on a per millimolar particle basis, respectively. These relaxivity values are much larger than the solid silica nanoparticles coated with multilayers of the Gd-DTPA derivative (see reference [23]). The enhanced magnetic resonance relaxivity was attributed to the ready access of water molecules through the nanochannels of mesoporous particles derivatized by gadolinium chelates Si-Gd-DTTA. When the functionalization is performed by a mixture of Si-Gd-DTTA and of rhodamine B conjugated to aminopropyltriethoxysilane (4 mol %, relative to Gd-Si-DTTA), the resulting mesoporous nanoparticles can behave as contrast agents for fluorescence and magnetic resonance imaging. The uptake of these fluorescent and paramagnetic mesoporous nanoparticles by monocyte cells was revealed both by the fluorescence and the contrast enhancement magnetic resonance images. Moreover, they appeared well suited for *in vivo* MRI after intravenous injection in the tail vein since aorta and liver are clearly visible on T_1- and

T_2-weighted images, respectively. These nanoparticles are indeed able to induce a positive and a negative contrast. If hepatic uptake renders liver imaging possible, thanks to the phagocytosis of the nanoparticles by the resident liver macrophage cells (Kupfer cells), it, however, leads to a quick clearance from the blood and limits the use of these nanoparticles as *in vivo* contrast agents.

Another interesting approach is to combine the incorporation of Gd with other lanthanide such as Eu and Tb since they are coordinated by the same ligands [38]. Eu^{3+} and Tb^{3+} possess excellent luminescent properties in the visible range. The post-functionalization of a fluorescent silica core, which contains Ru(bpy) as fluorophore with silylated polyaminocarboxylate ligands, allows the immobilization of both Tb^{3+} and Gd^{3+} ions. The presence of Tb^{3+} confers to the particles emission properties, which are complementary to those of Ru^{2+} entrapped inside the core, whereas gadolinium ions are expected to render the nanoparticles paramagnetic [39]. This approach enables the design of multimodal contrast agent that combine high resolution and sensitivity of optical set-up, with 3D localization provided by MRI.

The use of mesoporous nanoparticles as cell markers has also been studied. These structures are readily internalized by many type of cells, without the need of any transfection agent [28, 33, 36, 40]. Hsiao *et al.* have recently reported the use of hybrid mesoporous nanoparticles with Gd for stem cell tracking and were able to track as low as 4000 stem cells for 14 days using a 1.5T clinical MRI [34]. However, precise study on the effect of internalization on the contrast properties has to be performed for long-term cell trafficking visualization [41, 42].

4.2.4 Paramagnetic Quantum Dots

The functionalization of quantum dots (QD) by gadolinium chelates represents another route for the elaboration of fluorescent positive contrast agents combining FI and MRI.

QD refers to a crystallite cluster of II–VI semiconductors (CdS, CdSe, ZnS, CdSe@ZnS, etc.) exhibiting unique optical properties that arise from quantum confinement effects characteristic of these small dimensions (1–8 nm) [43–45]. Their attractive features (high photostability, high quantum yield, controllable and narrow emission band, broad excitation band) should contribute to a breakthrough of paramagnetic ions can be easily tethered to the polysiloxane shell provided that functional groups are present. In order to preserve the eight coordination sites of DOTA, an aromatic amine was grafted to one of the ethylene bridges of DOTA (NH_2-Bn-DOTA, Scheme 4.1). DOTA derivative with paramagnetic Gd^{3+} chelated are linked to the thiols of the silica shell embedding the QD by using the sulfo-SMCC crosslinker. The maleimide group of sulfo-SMCC allows the grafting to the thiol function due to the formation of thioether linkage while the NHS ester group at the opposite end of the crosslinker favors the formation of amide linkage with the aromatic amine of the DOTA derivative. As compared to unbound Gd-DOTA, the Gd^{3+} ion relaxivity r_1 for the gadolinium chelates attached to the silica-coated nanoparticles is increased by approximately six times (23 $mM^{-1}s^{-1}$ versus 3.7 $mM^{-1}s^{-1}$). Chemical analyses revealed that about 50 Gd-DOTA are grafted onto the nanoparticles. As a result, the paramagnetic QD (8 nm) exhibit a relatively high longitudinal relaxivity (1019 $mM^{-1}s^{-1}$, 1.4 T). Preliminary *in vivo* tests indicate that paramagnetic silica-coated nanoparticles after intravenous injection provide a contrast enhancement in MRI. The accumulation of paramagnetic nanoparticles in bladder was evidenced by the appearance of brighter zones in the bladder 5 min after the injection. However, these nanoparticles do not accumulate in the other organs since no contrast enhancement was observed elsewhere than the bladder. This indicates that nanoparticles freely circulate without undesirable accumulation and most of them are quickly removed by renal excretion.

Both previous studies led to the development of paramagnetic QD suitable for MRI applications. Their functionalization by gadolinium (III) chelates rests on the replacement of hydrophobic TOP/TOPO layer by hydrophilic one. However, the TOP/TOPO layer can be exploited for a further functionalization of QD as demonstrated by the studies performed by Bakalova *et al.* and by Mulder *et al.* [53–56]. The preservation of the TOP/TOPO layer limits the deleterious effect of polar species on the luminescence since the hydrophobic layer constitutes an efficient barrier.

In contrast to the procedure of Gerion *et al.*, which requires the replacement of TOP/TOPO layer by MPTMS [52], Bakalova *et al.* demonstrated that QD can be encapsulated by silica shell from hydrophobic layer [53]. Prior to the inorganic polymerization of the polysiloxane network, hydrophobic QD were covered by

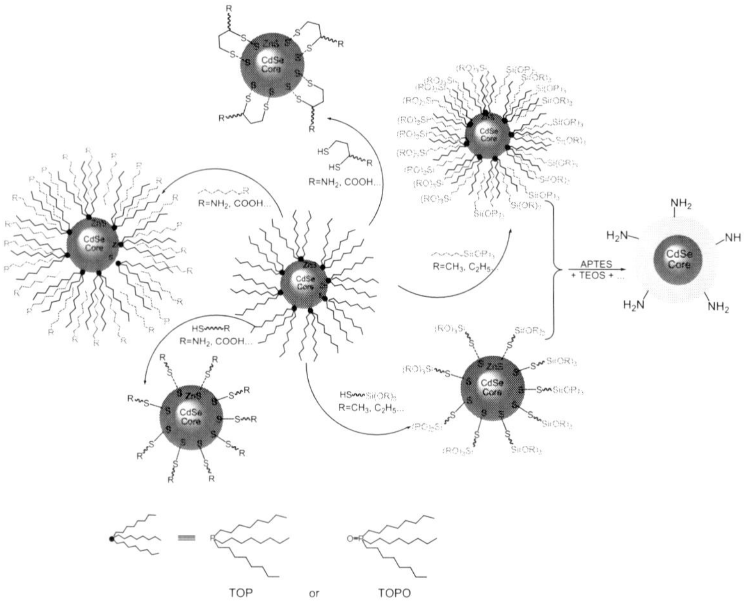

Scheme 4.7 Schematic of generic QD functionalization.

vinyltrimethoxysilane. The hydrophobic tail of this precursor (i.e., the vinyl moiety) was entrapped in the hydrophobic TOP/TOPO layer, whereas the methoxysilyl end group pointed towards outside. These methoxysilyl groups serve as anchoring sites for the polysiloxane shell yielded by hydrolysis — condensation of a mixture of TEOS and APTES. Owing to the amine function, which is a characteristic of the latter, surface is ready for surface modification. The MRI contrast agent ability was conferred to the QD either by the entrapment of amphilic gadolinium (III) chelates (trivial name: Resolve Al–Gd) before the encapsulation in the silica shell or by the surface modification of the silica-coated QD by hydrophilic gadolinium (III) chelates, which are DOTA derivatives (Gd-DOTA-Bn-NCS) (Scheme 4.1). The isothiocyanatobenzyl group (-CH_2-Φ-N=C=S) of Gd-DOTA-Bn-NCS carried by one of the ethylene bridges of the macrocycle ensures the covalent grafting of the gadolinium chelates by condensation with surface amine function leading to the formation of thiourea linkage (Schemes 4.4 and 4.5). Whatever the way for rendering them paramagnetic, silica-coated QD are able to induce the shortening of the longitudinal relaxation time T_1. This was reflected by the enhancement of the contrast of T_1-

weighted images. Various *in vitro* and *in vivo* studies showed that they can be applied as probes for multimodal imaging. Indeed the internalization of these particles can be monitored both by FI and MRI. It was demonstrated that silica-coated QD do not alter the viability of labeled cells. Furthermore, labeled cells under continuous UV or laser irradiation remain viable, whereas a cytotoxic effect is generated when cells are loaded with uncoated QD. The silica shell seems to suppress the photosensitization effect, which was previously reported for uncoated QD. The low toxicity of these contrast agents is encouraging for *in vivo* applications. However, the accumulation in spleen must be solved since this reflects the uptake by macrophages, which is commonly observed when positively charged particles are intravenously injected. This capture should be overcome by PEGylation.

The encapsulation of TOP/TOPO-coated QD by silica shell was also performed via reverse microemulsion technique [54]. In this case, the silica shell was formed between the QD and the TOP/TOPO layer. The post-functionalization of the resulting particles for rendering them paramagnetic and hydrophilic was realized according to the protocol developed by Mulder's research group for TOP/TOPO-coated QD [57].

Hydrophobic TOP/TOPO-coated QD can be rendered hydrophilic by coating them with amphiphilic PEGylated phospholipids (PEG-lipids: PEG-DSPE (1,2-distearoyl-sn-glycero-3-phosphoethanolamine-*N*-[methoxy-(poly(ethylene glycol))-2000]). These molecules are widely used to stabilize liposomes for pharmaceutical applications because PEG chains form a hydrophilic layer on the liposomal surface. The coating of hydrophobic QD (or QD encapsulated in silica shell) with a mixture of PEG-lipids and of paramagnetic lipids (i.e., gadolinium (III) chelate conjugated to hydrophobic stearylamide) yields water soluble and fluorescent positive contrast agents for FI and MRI. The post-functionalization of hydrophobic silica-coated QD with PEG-lipids improves markedly the biodistribution and the pharmacokinetics of these contrast agents [58].

Moreover, some PEG-lipids, which are terminated by maleimide moiety, allow for bioconjugation. Mulder's group succeeded in functionalizing these nanoprobes by RGD and annexin A5 proteins [55, 56]. The grafting of RGD, which is well known for targeting $\alpha_v\beta_5$ and $\alpha_v\beta_3$ integrins, favors the uptake of these paramagnetic QD by HUVEC due to the over-expression of $\alpha_v\beta_3$ integrins. The ability of these contrast agents to target $\alpha_v\beta_5$ and $\alpha_v\beta_3$ integrins opens the door to the detection of neo-angiogenosis, which is associated with the development of tumor, whereas the conjugation of Annexin A5 to these nanoprobes makes them well suited for the detection of apoptosis by FI and MRI. The strategy of the research group of Mulder, which is based on the synthesis of lipid-based nanoparticles, is very attractive because it leads to a large variety of contrast agents for multimodal imaging [57]. By a convenient choice of the building blocks (the lipids, the nanoparticles, the biotargeting groups, etc.), the behavior of these platforms can be accurately tuned for a definite application. Recently, they demonstrated that endogenous nanoparticles (high density lipoproteins, HDL) can be also modified for multimodal imaging by incorporating paramagnetic and/or fluorescent lipids but also by including within each HDL nanoparticle hydrophobic inorganic core such as iron oxide, QD, or gold [59]. Since it is known to migrate to atherosclerotic plaques and remove cholesterol from macrophages, HDL was exploited by Frias *et al.* to develop a probe for molecular imaging of macrophages [60, 61]. The inclusion of inorganic core affords additional features for imaging purposes. For instance, the presence of a hydrophobic gold core in a paramagnetic and fluorescent HDL allows monitoring gold-HDL by MRI, FI and X-ray imaging since the high atomic number of gold element confers to the gold nanoparticles the ability to absorb X-ray photons.

4.2.5 Paramagnetic Gold Nanoparticles

One of the most promising applications of nanosizes gold objects lies in their ability to combine imaging and therapy [62]. Two main strategies were developed for designing gold nanoparticles that can be used for imaging and destroying tumors. The first one results from the properties of gold element, whereas the second one rests on the tunable optical properties of gold particles. Gold particles are indeed characterized by strong light diffusion and absorption cross sections, which are sensitive to the size, shape, and dielectric environment [63]. By tuning these parameters, gold-based nanomaterials (gold-coated silica particles and gold nanorods) can be synthesized and exploited for NIR imaging and therapy as demonstrated by Halas *et al.* and El-Sayed *et al.* [64, 65]. Whatever the size, shape, and environment of the gold nanomaterials, they strongly absorb X-ray photons. Hainfeld *et al.* demonstrated that ultrasmall hydrophilic gold nanoparticles (1.9 nm) can be monitored by X-ray imaging after intravenous injection to mice and applied for radiotherapy [66, 67].

However, X-ray imaging is not completely harmless for envisaging a long-term biodistribution study without interference with radiotherapy. Moreover, this technique suffers a low sensitivity. In order to circumvent this problem, the use of gold nanoparticles designed for MRI was proposed [68, 69]. Roux *et al.* succeeded in designing gold nanoparticles for MRI by reducing gold salt in presence of a dithiolated ligand DTDTPA (Au@DTDTPA). As a result, 2.4 nm sized gold cores embedded in a DTDTPA shell were yielded. DTDTPA is a dithiolated derivative of DTPA, which is commonly used as gadolinium chelate (DTPA-Gd) for clinical MRI. The transformation of DTPA into DTDTPA was performed by the condensation of a DTPA bis-anhydride molecule with two aminoethanethiol molecules (Fig. 4.1). It was demonstrated that each gold core was encapsulated in a multilayered shell, which was composed of about 150 DTDTPA molecules. The tethering and formation of this multilayered shell are actually ensured by the presence of two thiol functions per DTDTPA. Since thiol function exhibits a great affinity for gold, some DTDTPA molecules present in the shell are immobilized onto gold nanoparticles through the strong interaction between gold and sulfur atoms of one thiol function. Thiol functions not involved in the grafting of the shell onto the nanoparticles were implied in the formation of disulfide bonds. These bonds render the formation of multilayers around each gold core possible. Due to the presence of DTDTPA shell, until 150 Gd^{3+} ions can be immobilized on a gold nanoparticle (Au@DTDTPA-Gd_{150}). However, a good compromise between colloidal stability and amount of active ions for MRI was found when the number of gadolinium (III) per particle is fixed at 50 (Au@DTDTPA-Gd_{50}). The behavior of Au@DTDTPA-Gd_{50} in the bloodstream after intravenous injection can be easily monitored by both X-ray imaging (in transmission and computed tomography modes) and MRI (r_1 = 3.9 $mM^{-1}s^{-1}$ based on gadolinium concentration and r_1 = 195 $mM^{-1}s^{-1}$ based on nanoparticle concentration) (Fig. 4.1). Biodistribution studies performed with both medical imaging techniques and with post-mortem elemental analysis of organs revealed that these paramagnetic gold nanoparticles freely circulate in the bloodstream and do not accumulate in liver, spleen, and lungs. In the case of tumor-bearing rats (or mice), it must be, however, pointed out that accumulation was observed in tumor due to the EPR effect. This phenomenon was exploited for radiotherapy [70]. It was demonstrated that the survival of diseased rats was lengthened only when they were treated by X-ray beam after intravenous injection of Au@DTDTPA-Gd_{50}. Due to their imaging and therapeutic potentialities, Au@DTDTPA-Gd_{50} seems suitable for *in vivo* medical applications.

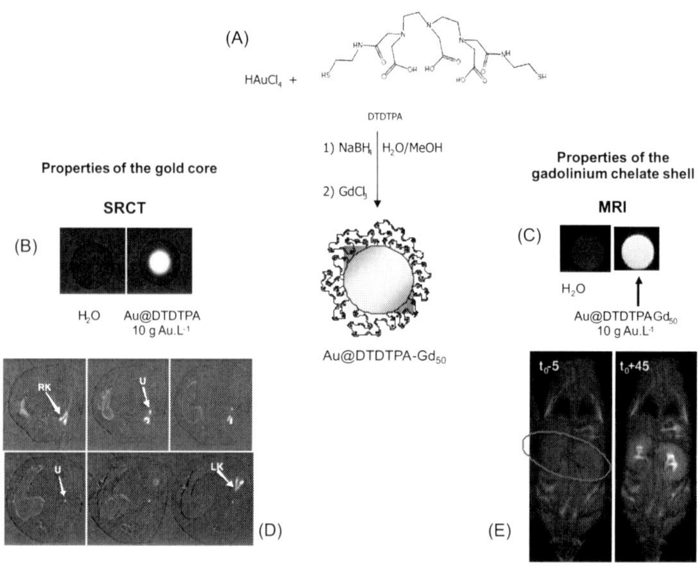

Figure 4.1 (A) Synthesis of Au@DTDTPA-Gd_{50}, (B) Synchrotron radiation computed tomography (SRCT), and (C) Magnetic resonance images of phantoms containing water and aqueous suspension of Au@DTDTPA-Gd_{50} ([Au] = 10 gL^{-1}). (D) SRCT images of a series of successive transverse slices, including right kidney of a rat from the head (top left) to the tail (right down). The images were recorded 10 min after the intravenous injection of Au@DTDTPA-Gd_{50} to a rat (RK and LK for right and left kidneys, respectively, U for ureter). (E) T_1-weighted images of a mouse 5 min before (t_0 – 5) min and 45 min after intravenous injection of Au@DTDTPA-Gd_{50} (K for kidneys and B for bladder).

Gold nanoparticles coated with rare earth chelates were also produced for the development of fluorescent sensor since some rare earth ions are characterized by a very photostable luminescence, which distinguished from organic dyes and QD by narrower emission peaks and longer lifetime of excited state. Gunlaugsson *et al.* demonstrated that the luminescence of europium buried into a heptadendate ligand (a tetraazamacrocycle) can be modulated by exploiting the deleterious quenching effect of water on the rare earth luminescence [71]. The luminescence is switched off when two water molecules are present in the coordination sphere, whereas the displacement of water by a diketone derivative bearing an aromatic ring re-establishes the light, which is even enhanced due to the effect of the antenna from aromatic ring. This principle developed for modulating fluorescence can be applied for "smart" MRI. When water is present in the coordination sphere, these gold particles can exert an enhancement of the positive contrast of magnetic resonance images, whereas no enhancement is visible when water was displaced by analytes. Therefore, MRI contrast agents, which are sensitive to the presence of definite biomolecules, can be imagined.

Gold nanoparticles can also be rendered paramagnetic by the covalent grafting of gadolinium (III) chelates to biomolecules, which are then immobilized on the particles for targeting them to the zone of interest. For instance, Chung *et al.* applied this strategy for preparing paramagnetic gold nanostructures, which are able to combine dual modality imaging (MRI and optical imaging) and targeted therapy [72]. First, hollow gold nanoparticles coated with thiol-ended PEG and cystein-protein-G were synthesized. Each class of molecules grafted to the hollow nanoparticles ensures a specific role. PEG was used for improving the colloidal stability and, therefore, the particle blood circulation time and the efficiency of particle internalization, while protein G molecules were introduced to optimize the orientation of antibodies (anti-HER2) on the surface of the hollow gold nanoparticles. Meanwhile, anti-HER2 antibodies were derivatized by gadolinium chelates by exploiting the covalent coupling between the amino group of the antibody and the isothiocyanate group of p-SCN-Bn-DTPA (Schemes 4.1 and 4.5). After chelation of gadolinium (III) ions by DTPA conjugated to anti-HER2, the paramagnetic antibody (anti-HER2-Gd-DTPA) was immobilized onto hollow gold nanoparticles. The resulting particles were successfully applied as both MRI (r_1 = 23.7 $mM^{-1}s^{-1}$) and optical imaging contrast enhancers of cancer cells since they can be observed by optical microscopy (due to the optical properties of hollow gold nanoparticles) and by MRI. The presence of anti-HER2 onto the nanoparticles favors the internalization in SKBR3 cells (breast cancer cells, which overexpress epidermal growth factor receptors targeted by anti-HER2). Labeled cells, which can be, therefore, visualized by dark-field-based scattering images and MRI, were destroyed after NIR irradiation due to the photothermal effect, which occurs when NIR light is absorbed by gold nanostructures. The multifunctional hollow gold nanoparticles are also expected to be used for imaging specific cancer sites using MRI, and then for destroying cancer cells selectively via illumination with NIR light. However, their large size should impede the *in vivo* application of these paramagnetic therapeutic agents. For this reason, more information regarding their biodistribution and pharmacokinetic parameters, which are not available, are required.

4.3 Gadolinium (III) Containing Crystalline Nanoparticles

4.3.1 Introduction

The development of gadolinium (III)-based nanoparticles, which is recounted in this chapter, aims at affording positive contrast agents for MRI with a stronger ability of contrast enhancement (due to both the increase in gadolinium (III) number and the decrease in rotational motion), better pharmacokinetic parameters, a better control of biodistribution and a larger palette of properties than that of molecular gadolinium complexes. To achieve this goal, Gd-DTPA or -DOTA functionalized polymer [73, 74], self-assembled peptide amphiphiles [75, 76] or viral capsid [77], Gd-DTPA-terminated dendrimer [78, 79], gadolinium complexes-loaded liposomes [80], high-density lipoprotein nanoparticles [60, 61], micelles [81, 82], or polymeric nanoparticles [83, 84], gadolinium ions entrapped in zeolites [85], fullerenes [86–89], carbon nanotubes [90, 91], clays [92] or mesoporous silica nanoparticles [38], and gadolinium chelates immobilized on quantum dots [55, 56], on lipid particles [93], and on gold nanoparticles [68, 69] were synthesized and studied. They are all characterized by an increase in molecular weight and of the amount of Gd(III) ions per contrast agent. As a consequence of their structure, some of them were easily functionalized by biotargeting groups and/or fluorescent molecules, conferring them additional attractive features [55, 56, 60, 61, 73, 77–80, 93]. Although the number of gadolinium atoms is very high even for smaller size, the potential of crystalline nanoparticles based on inorganic gadolinium such as gadolinium oxide [94–96], gadolinium fluoride nanoparticles [97], and gadolinium carbonate [98, 99] has been only recently evaluated.

Pioneer studies devoted to gadolinium oxide nanoparticles were performed by Roberts and Watkins [95, 96]. They revealed the potential of gadolinium oxide nanoparticles to be applied as contrast agents for MRI. However, these preliminary works suffered from the availability of well-defined hydrophilic gadolinium oxide nanoparticles. The exploitation of the attractive characteristics of gadolinium oxide nanoparticles for MRI became conceivable only when a reproducible and efficient synthesis protocol of naked gadolinium oxide nanoparticles and their functionalization were developed.

4.3.2 Synthesis and Functionalization of Gadolinium Oxide Nanoparticles

The research activity of Tillement's group was decisive in this field. First, a reproducible and reliable synthesis of gadolinium oxide nanoparticles was reported by this group [100, 101]. These particles were synthesized by applying, with some modifications, the polyol route, which opened the door to the synthesis of a large range of inorganic crystalline nanoparticles. The synthesis of gadolinium oxide nanoparticles is based on the alkaline hydrolysis of gadolinium chloride in diethylene glycol (DEG), a high boiling diol. Diethylene glycol plays an important role. It enables the synthesis at 180°C and to accurately control the size of gadolinium oxide nanoparticles (between 1 and 5 nm according to the experimental conditions) because the high viscosity of DEG and its adsorption on growing particles limit their growth and avoid the agglomeration. The alkaline hydrolysis of a mixture of gadolinium chloride and of chloride salt of luminescent rare earth ions (Tb^{3+}, Eu^{3+}, Nd^{3+}, ...) in boiling DEG provides rare earth (RE) doped gadolinium oxide nanoparticles whose Gd to RE molar ratio is very similar to the initial ratio [100, 101]. Since gadolinium oxide is transparent in the visible range of light, it was demonstrated that these nanoparticles are fluorescent when rare earth ions are present in the crystalline nanoparticles. Despite the remarkable optical properties of luminescent rare earth ions (high photostability, narrow emission peaks, long lifetime of excited state), the *in vivo* application of luminescent-doped gadolinium oxide nanoparticles is impeded by the low quantum yield and by the energy of the excitation wavelength (UV range). However, the *in vivo* application of gadolinium oxide nanoparticles remains very promising for MRI provided that the surface of gadolinium oxide core is adapted to biological environment. The second important contribution of Tillement's research group lies in the functionalization of these cores. The strategy developed by this group rests on the encapsulation of the gadolinium oxide cores in a polysiloxane shell [102]. This functionalization mode provides several important advantages. Besides a protective effect of the core, which was validated by the preservation of the luminescence in acidic media when the cores are doped by Tb^{3+} ions, the polysiloxane shell exerts a positive influence on the luminescence of rare earth ions present in the cores since its presence induces the replacement of RE-OH bonds, which behaves as quenchers for the luminescence by RE-O-Si bonds and an energy transfer from polysiloxane shell to luminescent centers [103]. The encapsulation of luminescent rare earth doped gadolinium oxide cores yielded nanomaterials, which are more fluorescent than the uncoated gadolinium oxide cores are.

But the most striking feature of the polysiloxane shell results from its chemical nature, which can provide new properties to the particles. The polysiloxane shell was indeed obtained by hydrolysis-condensation of tetraethyl orthosilicate (TEOS) and by aminopropyltriethoxysilane (APTES) (see Section 4.2.2). The use of APTES allows the sequential functionalization of the polysiloxane shell: before and after the hydrolysis-condensation step, i.e., before and after the formation of the polysiloxane shell because each APTES carries an amino (NH_2) function. Before the hydrolysis condensation, APTES can be conjugated to organic dyes thanks to the coupling between the amino group of APTES and the isothiocyanate or NHS ester group of the organic fluorophores, while the polysiloxane shell can be post-functionalized by the covalent grafting of hydrophilic PEG chains [104, 105] and/or of biotargeting groups [102, 106]. The encapsulation of gadolinium oxide cores widens the properties range of these particles because the inner part of the polysiloxane shell is functionalized by organic dyes and the outer part is derivatized by hydrophilic molecules.

As expected, these gadolinium oxide nanoparticles encapsulated in fluorescent and PEGylated polysiloxane shell (GadoSiPEG) can be followed up after intravenous injection to mice and rats by fluorescence imaging (due to Cyanine 5 covalently bound to the polysiloxane network) and by MRI (thanks to the presence of gadolinium (III) in the core of each particle) (Fig. 4.2). The *in vivo* imaging experiments revealed that the particles exhibit attractive biodistribution and pharmacokinetics parameters, which are accurately controlled by the length of the PEG chains and the nature of the end group [104, 105]. They freely circulate in the bloodstream without undesirable nonspecific accumulation in liver, spleen, and lungs. Moreover, these nanoparticles are rather quickly removed from body essentially by renal excretion (Fig. 4.2). All these observations denote a safe behavior of GadoSiPEG nanoparticles when they are intravenously injected to small animals. However, the accumulation of these nanoparticles was observed after intravenous injection to gliosarcoma (9L)-bearing rats. This was revealed by the obvious delineation of tumor due to the enhancement of the positive contrast induced by the presence of the particles in the tumor region. This accumulation, which can be explained by the EPR effect, was exploited for radiotherapy since the atomic number of gadolinium element is sufficiently high for generating a dose enhancement of X-ray. The treatment of brain tumor-bearing rats by radiotherapy led to a longer survival only when hybrid gadolinium oxide nanoparticles were administered by intravenous injection before the exposure to therapeutic X-ray beam [107, 108]. The therapeutic activity of gadolinium oxide nanoparticles is not limited to the radiotherapy. As widely mentioned in the literature, objects containing gadolinium carry the promise to replace boron-based compounds for neutron capture therapy. The internalization of GadoSiPEG nanoparticles in bioluminescent murine lymphoma cells (EL4-luc), which was monitored by fluorescence (Rhodamine B isothiocyanate was covalently bound to the polysiloxane shell), MRI and elemental analysis (ICP) generates no cytotoxicity and no alteration of the cell proliferation until $[Gd]_{incubation} = 0.3$ mM. Unloaded cells are also not affected by the exposure to a thermal neutron beam if the delivered dose is equal to 3Gy (or less). But the irradiation of EL4-luc cells after incubation in presence of GadoSiPEG ($[Gd]_{incubation} = 0.05$ mM) by harmless thermal neutron beam (3 Gy) generates a great killing effect since all cells are destroyed [109]. Gadolinium oxide cores embedded in a fluorescent and PEGylated polysiloxane shell are very attractive multifunctional nanostructures for *in vivo* application and in peculiar for imaging guided therapy since they combine FI (organic dyes conjugated to the inner part of the polysiloxane shell), MRI (gadolinium (III) ions in the crystalline core), and therapeutic activity, which exploits some properties of gadolinium element (radio and neutrontherapy).

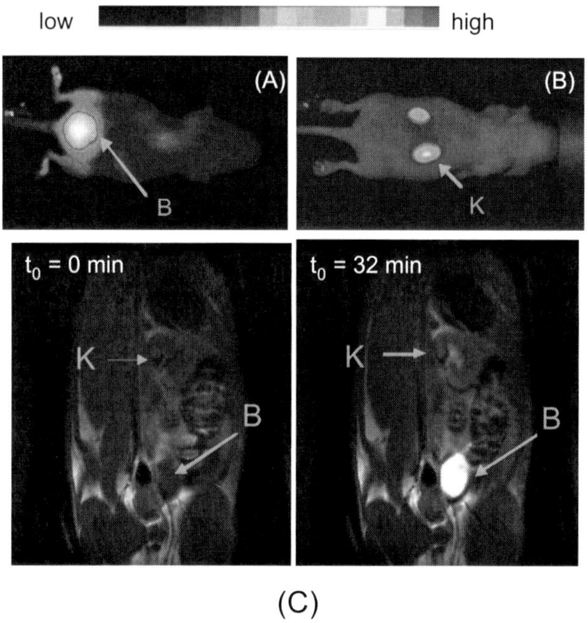

Figure 4.2 Fluorescence reflectance imaging of a nude mouse lain on back (A) and on belly (B) 15 min after the intravenous injection in caudal vein of gadolinium oxide nanoparticles embedded in a fluorescent (Cy5 dye) and PEGylated polysiloxane shell (GadoSiPEG) (exposure time: 200 ms). (C) T_1-weighted images of rat before (t_0 – 4) min and 32 min after intravenous injection of hybrid gadolinium oxide nanoparticles embedded in a fluorescent and hydrophilic polysiloxane shell in caudal vein (K and B for kidney and bladder, respectively).

The potential of gadolinium oxide cores for MRI was confirmed by a research group of Linköping University (Sweden) [94]. Their works are interesting and complementary to the one done by Tillement's group. Indeed, they propose an alternative way for derivatizing the gadolinium oxide cores and, therefore, for exploiting their paramagnetic character in MRI experiments. Their strategy consists of capping of gadolinium oxide nanoparticles, which were synthesized according to the polyol route [100, 101] by organic acids or by the grafting of PEG-silane molecules. After their synthesis in DEG, the resulting gadolinium oxide cores are coated by DEG molecules, which can be replaced in hot DEG by citric acid or dimercaptosuccinic acid [110, 111]. This layer of organic acid molecules acts as a primer for further functionalization and, in peculiar, for PEGylation, which is required for biological applications. The grafting of nonimmunogenic and nonantigenic PEG is expected to render the gadolinium oxide cores more robust, to improve the colloidal stability in biological media, to favor uptake by cells and enhance blood retention. Citric acid-coated gadolinium oxide cores were derivatized by PEG chains ended by a thiol (SH) group and an amino (NH_2) group (Gd_2O_3@CA-PEG). The amino group of PEG permits the condensation with carboxylic acid groups of citric acid, which were activated by EDC/NHS mixture prior to the coupling reaction. Thiol groups at the opposite extremity can be used for the grafting of molecules bearing maleimide group [112]. Since each dimercaptosuccinimide (DMSA) molecule grafted to gadolinium oxide cores carries two thiol functions, Mal-PEG-NHS (Mal and NHS for maleimide and *N*-hydroxysuccinimide groups, respectively) was used for the PEGylation of DMSA-capped gadolinium oxide cores. The covalent coupling of PEG is ensured by the formation of thioether linkage resulting from the condensation of the thiol group of DMSA with maleimide moiety of Mal-PEG-NHS (Gd_2O_3@DMSA-PEG) [112]. The presence of NHS group affords the possibility of grafting aminated molecules (typically biomolecules, which are rich in amine functions) onto the PEGylated nanoparticles [112]. The PEGylation can also be performed using PEG-silane since alkoxysilyl end-group of PEG-silane favors its grafting on oxide nanoparticles through the condensation between the alkoxysilyl groups and hydroxyl (OH) groups present on the surface of the oxide materials (Gd_2O_3@PEG) [113]. These PEGylated oxide nanoparticles, which are prepared by the research group of Linköping University, exhibit, as expected, relatively high longitudinal relaxivity r_1. The relaxivity values of Gd_2O_3@PEG, Gd_2O_3@DMSA-PEG, and Gd_2O_3@CA-PEG (based on the amount of gadolinium (III) ions in each sample) were 12.0, 14.2, and 7.8 $mM^{-1}s^{-1}$, respectively [114]. Except for Gd_2O_3@CA-PEG, these PEGylated gadolinium oxide nanoparticles exhibit a higher relaxivity than the one of GadoSiPEG developed by Tillement's research group (see above). The longitudinal relaxivities of GadoSiPEG nanoparticles containing the smallest gadolinium oxide cores (2.2 nm and 3.8 nm) were 8.8 $mM^{-1}s^{-1}$ whatever the core size, whereas r_1 dropped to 4.4 $mM^{-1}s^{-1}$ when the size of the core amounted to 4.6 nm [104]. This difference can be explained by a better accessibility of water molecules to the surface of the oxide cores in case of Gd_2O_3@PEG, Gd_2O_3@DMSA-PEG because the cores are encapsulated in a monolayer, whereas the gadolinium oxide cores of GadoSiPEG are embedded in a relatively thick polysiloxane layer. However, the cause of this difference is probably not so obvious since Gd_2O_3@CA-PEG particles exhibit a lower relaxivity than GadoSiPEG do, despite the thin PEG shell of Gd_2O_3@CA-PEG. It must, however, be pointed out that two different strategies for the functionalization of gadolinium oxide cores are available. They provide similar performance in terms of positive contrast enhancement for MRI. Unfortunately, their behavior after intravenous injection in small animals (rats, mice) cannot be compared because no *in vivo* imaging experiment was reported with the PEGylated organic acid capped gadolinium oxide cores in contrast to GadoSiPEG. Although the encapsulation of the gadolinium oxide cores by polysiloxane shell seems more difficult to control, it allows, nevertheless, enlarging more easily the palette of properties of the contrast agents since post-functionalization of gadolinium oxide cores can be sequentially realized, in contrast to the encapsulation by organic acids. Besides MRI imaging, which is efficient whatever the functionalization mode, GadoSiPEG nanoparticles can indeed be easily followed up by FI after their uptake by cells or after their intravenous injection in small animals [104, 105, 109, 115], whereas the detection of PEGylated organic acid-capped gadolinium oxide nanoparticles by FI rests only on the doping of the core by rare earth ions [114]. Despite its interesting properties, the FI based on the luminescence of rare earth ions (Tb^{3+}) is not very well suited for biological applications because the light emission of Tb^{3+} in gadolinium oxide is induced by the absorption of UV photons, which are harmful for living matter. Moreover, the UV excitation is not suited for FI of small animals since the UV irradiation is almost totally absorbed by the skin.

Because of the great chemical similarities of rare earth elements, the synthesis of gadolinium oxide nanoparticles by applying the polyol route was easily adapted to the synthesis of dysprosium oxide nanoparticles (diameter: 1 nm). It was demonstrated by Peters *et al.* that these particles exhibit a high transverse relaxivity r_2, which is dependent on the size [116]. The optimal size for the highest r_2

value is between 60 and 70 nm: r_2 varies between 190 $mM^{-1}s^{-1}$ (70 nm at 7 T) and 675 $mM^{-1}s^{-1}$ (60 nm at 17.6 T). Large 1 H transverse relaxation rate enhancements in aqueous suspensions are also observed for other paramagnetic Ln_2O_3 particles (with Ln = Gd, Er) [117]. Although these particles show paramagnetic behavior, the magnitude of this effect is comparable with the effect caused by superparamagnetic iron oxide particles [118] but only at high external magnetic field. Since there is a tendency to perform MRI exams at higher magnetic fields, lanthanide oxide nanoparticles have favorable relaxivity properties for these higher fields.

Other routes for the preparation of gadolinium oxide based contrast agents have been recently developed. Gadolinium oxide nanoparticles in carbon nanotubes were synthesized for MRI application [119]. They were obtained after thermal treatment of gadolinium acetate entrapped in single-wall carbon nanohorns (SWNHs) with holes (NHox). Gd_2O_3-NHox dispersed in agarose gel induces a significant positive contrast enhancement as revealed by T_1-weighted images of the phantoms. However, no r_1 value was given probably because the variation of $1/T_1$ as a function of gadolinium concentration is not linear. Although the enhancement induced by Gd_2O_3-NHox appears sufficient for MRI, their application as *in vivo* contrast agent seems to be limited due to the lack of colloidal stability in biological fluid and the toxicity of carbon nanotubes.

Since the colloidal stability of paramagnetic nanoparticles is a prerequisite for their application as *in vivo* contrast agent for MRI, the encapsulation of gadolinium oxide nanoparticles (5 nm) in the protein apoferritin has been proposed [120]. The protein shell prevents bulk aggregation of the particles, rendering them water soluble. These nanoparticles exhibit longitudinal and transverse relaxivities higher than those of classic Gd(III)-complexes (10–25 and 70 times, respectively). Moreover, the r_2/r_1 ratio is drastically altered by changing the frequency, thus giving the possibility to tune the response in terms of image contrast. The external apoferritin shell also affords the opportunity of covalently grafting organic molecules for incorporating new functionality or for increasing the plasma half-life. However, no data on *in vivo* imaging experiments are available when this chapter was written.

Yeh *et al.* succeeded in synthesizing porous and hollow gadolinium oxide particles [121]. They demonstrated that hollow spheres produced better positive contrast enhancement (r_1 = 17.7

$mM^{-1}s^{-1}$ and r_2/r_1 = 1.50) than porous particles did (r_1 = 16.8 $mM^{-1}s^{-1}$ and r_2/r_1 = 1.87). Such materials are very interesting from the chemistry point of view; however, their application as contrast agents for MRI can be impeded by the large size of these particles. The size of hollow Gd_2O_3 spheres (~130 nm) and that of porous Gd_2O_3 particles (~187 nm) are obviously larger than the hydrodynamic diameter recommended for the renal excretion of nanoparticles in the bloodstream (D_h < 5 nm) according to the studies of Frangioni *et al.* [49].

4.3.3 Synthesis of Gadolinium-Containing Crystalline Nanoparticles

At the same time, several routes were explored for obtaining other types of crystalline nanomaterials rich in gadolinium (III) by wet chemistry techniques. The potential of these gadolinium-containing crystalline nanoparticles for MRI was naturally evaluated since there is a real need for efficient multifunctional contrast agents. Among the variety of materials, fluoride and carbonate particles have received much attention.

4.3.3.1 Fluoride nanoparticles containing gadolinium (III)

Van Veggel *et al.* acquired a wealth of experiences in the synthesis of lanthanide fluoride nanoparticles for luminescence application. They turned a peculiar attention to the synthesis and characterization of gadolinium fluoride. The synthesis of nanoparticles were carried out for 3–4 h in water at moderate temperature (75°C) and rests on the reaction of ammonium hydroxide (NH_4OH) with sodium fluoride and gadolinium (III) nitrate ($Gd(NO_3)_3{\cdot}6H_2O$) in the presence of hydrophilic stabilizing agent (citric acid or 2-aminoethylphosphate (AEP)). As a result, relatively large gadolinium fluoride nanoparticles capped by citric acid or AEP (GdF_3@cit (mean diameter 129.3 nm) and GdF_3@AEP (mean diameter 51.5 nm), respectively) are collected [97]. The presence of hydrophilic agents on the surface of the particles facilitates their dispersion in aqueous solution and prevents them from agglomeration. Moreover, citric acid and especially AEP allow the post-functionalization of the nanoparticles by biotargeting group (biotin) and by PEG chains [122]. Whatever the capping ligands, the GdF_3 nanoparticles exhibit moderate longitudinal relaxivities (3.17 and 2.71 $mM^{-1}s^{-1}$ per Gd^{3+} for GdF_3@cit and GdF_3@AEP, respectively) when it was expressed in terms of the number of gadolinium ions in the nanoparticles. But the relaxivities per mM nanoparticles reach high values, which amount to 2.0×10^7 and 8.8×10^5 $mM^{-1}s^{-1}$ for the GdF_3@cit and GdF_3@AEP nanoparticles, respectively. This study suggests that Gd^{3+} ions located beneath the outer shell of Gd^{3+} are also contributing to the relaxation. This was confirmed by the works focused on GadoSiPEG (see Section 4.3.2), which showed that the effect on r_1 does not only result from the Gd^{3+} ions present at the surface of the crystalline core but also partially from those which are inside the core. Indeed the relaxivities r_1 are identical for 2.2 and 3.8 nm Gd_2O_3 cores, whereas the smallest particles possess the highest amount of surface Gd^{3+} ions for a same gadolinium concentration. It means that in both samples, all Gd^{3+} ions have the same influence whatever their location in the crystalline structure and that the range of the influence of Gd^{3+} ions is at least of 1.9 nm. The presence of citric acid and AEP molecules onto the GdF_3 nanoparticles (GdF_3@ cit and GdF_3@AEP, respectively) allows grafting of highly hydrophilic molecules and biotargeting groups (e.g., biotin) for controlling solubility and retention in specific tissues [122]. Although very high chemical stability, which is reflected by a very low leaching of ions over considerable periods of time, renders these gadolinium fluoride nanoparticles very attractive, their potential as positive contrast agents for MRI must be confirmed by *in vivo* experiments.

$NaYF_4$ nanocrystals constitute another interesting class of fluoride materials since they can be easily doped by various rare earth ions. These fluoride nanocrystals received much attention for photoluminescence applications because they are considered as the most efficient host materials for upconversion phosphors. They exhibit visible emission upon NIR excitation when doped with Yb^{3+} and Er^{3+} ions. By an appropriate choice of rare earth ions incorporated in the fluoride matrix, the emission wavelength can be red-shifted relatively to the excitation wavelength. Downconversion is, therefore, achieved and can be exploited for biological imaging. Chatterjee *et al.* have recently demonstrated the use of upconversion nanophosphors ($NaYF_4$: Er^{3+}, Yb^{3+}) for *in vitro* imaging of cancer cells and *in vivo* imaging in tissues [123]. It is claimed that the advantages of these nanoprobes include absence of photodamage to living organisms, very low autofluorescence, high detection sensitivity, and high light penetration depth in biological tissues. However, due to the low penetration depth of visible light in tissues, *in vivo* imaging based on luminescent nanoprobes showing NIR-to-visible upconversion is still limited. Prasad *et al.* and Li *et al.* demonstrated that *in vivo* imaging of whole-body small animals is possible when upconversion nanocrystals based on $NaYF_4$ are co-doped by Yb^{3+} and Tm^{3+} since the presence of Tm^{3+} allows NIR-to-NIR upconversion [124, 125].

Despite many studies devoted to the photoluminescence of rare earth doped $NaYF_4$ nanocrystals, reports on the doping of these crystals by gadolinium are rare despite the great potential of contrast agents with multimodality imaging ability. Prasad *et al.* contributed to the development of paramagnetic fluoride nanophosphors of size less than 50 nm [126]. They reported the synthesis of functionalized fluoride nanocrystals, which form stable aqueous dispersion, which can be used for both optical imaging and MRI. Down- and up-converted luminescence from the rare earth ions (Eu^{3+}, Er^{3+}, and Yb^{3+}) doped into nanometer-sized fluoride matrices allows optical imaging modality, and co-doping with Gd^{3+} confers the resulting nanoparticles the ability to enhance the positive contrast of magnetic resonance images. By a convenient surface derivatization (immobilization of antibodies (anti-claudin and anti-mesothelin) via the covalent coupling with mercaptopropionic acid anchored onto the nanoparticles), these nanoparticles can be selectively taken up by pancreatic cancer cells. The selective internalization in cells was demonstrated only by the light emitted by the nanoparticles (using a confocal microscope). Since the value of the longitudinal relaxivity r_1 of these Gd^{3+} co-doped fluoride nanocrystals is very low (r_1 = 0.14 $mM^{-1}s^{-1}$ and r_2/r_1 = 62 >> 1), in comparison to gadolinium oxide or gadolinium fluoride (GdF_3) nanoparticles, the uptake of these nanoprobes cannot be monitored by MRI. The low value of r_1 observed in case of Gd^{3+} co-doped $NaYF_4$ can be explained by the amount of gadolinium (III) ions in the crystalline fluoride matrix, which is largely lower than in case of gadolinium oxide or gadolinium fluoride (GdF_3) nanoparticles. If a small amount of luminescent rare earth ions is sufficient for detection by optical imaging techniques, this study showed clearly that a greater amount of paramagnetic ions is required for inducing positive contrast enhancement in MRI because this medical imaging technique is less sensitive. To overcome this limitation, Li *et al.* proposed the synthesis of hydrophilic $Tm^{3+}/Er^{3+}/Yb^{3+}$ co-doped $NaGdF_4$ nanoparticles [127]. As expected, these particles whose size lies in the range of 25–60 nm exhibit NIR-to-NIR upconversion due to the presence of doping species ($Tm^{3+}/Er^{3+}/Yb^{3+}$). Moreover, the replacement of Gd-doped $NaYF_4$ by $NaGdF_4$ leads to luminescent nanoparticles with higher longitudinal relaxivity r_1 (r_1 = 5.60 $mM^{-1}s^{-1}$ vs r_1 = 0.14 $mM^{-1}s^{-1}$). This improvement is attributed to a higher Gd^{3+} concentration on the surface of the $Tm^{3+}/Er^{3+}/Yb^{3+}$ co-doped $NaGdF_4$ nanoparticles as compared to Gd-doped $NaYF_4$ nanocrystals. The authors demonstrated that these nanoparticles meet the criteria for *in vivo* imaging because they are characterized by good water-solubility and low cytotoxicity, besides their attractive optical and magnetic properties. After intravenous injection, the hydrophilic $Tm^{3+}/Er^{3+}/Yb^{3+}$ co-doped $NaGdF_4$ nanoparticles can be followed up both by fluorescence imaging and by MRI. The *in vivo* and *ex vivo* imaging experiments revealed that these nanoparticles accumulated only in liver and spleen. Nonspecific accumulation in these organs is generally ascribed to uptake by resident phagocytes in liver (Kupfer cells) and spleen (macrophages and B cells). Although this feature can be exploited for imaging liver and spleen, such nonspecific accumulation is generally undesirable because it induces a too quick removal of a large part of contrast agents from the bloodstream. This can be detrimental for the monitoring of passive or active targeting.

4.3.3.2 Carbonate particles containing gadolinium (III)

By refluxing aqueous solution containing urea and gadolinium chloride, amorphous gadolinium carbonate ($Gd_2O(CO_3)_2{\cdot}H_2O$) particles can be obtained [99]. The shape (spherical, rhombus- or rice-shaped nanoparticles) and the size can be tuned by varying the urea to gadolinium chloride molar ratio. Relaxation rate measurements revealed that the relaxivities depend on the size and shape of the particles. In case of 500 nm sized spherical $Gd_2O(CO_3)_2{\cdot}H_2O$ particles, r_1 is equal to 16.5 $mM^{-1}s^{-1}$ (r_2/r_1 = 12.7, 3T). After injection, the behavior of these large particles can be monitored by MRI. As expected for large particles, accumulation in liver was observed.

These particles can be functionalized by an aminated polysiloxane shell due to the hydrolysis-condensation of the aforementioned mixture of TEOS and APTES (Scheme 4.3). This polysiloxane shell permits a further functionalization (Scheme 4.5). Due to the presence of amino groups onto the nanoparticles, gold nanoparticles can be immobilized since nitrogen atoms have a great affinity to gold atoms. These gold nanoparticles act as seeds for the growth of a gold layer. $Gd_2O(CO_3)_2 \cdot H_2O$ particles can also be encapsulated by a gold shell whose thickness can be accurately controlled [98]. The encapsulation of these gadolinium carbonate particles by a gold shell induces a decrease in the longitudinal relaxivity with the increase in the thickness of the gold shell. However, the gold shell renders the resulting nanoparticles suited for photothermal therapy [64]. $GdO(CO_3)_2 \cdot H_2O$ particles encapsulated in gold shell are able to combine MRI and therapy as revealed by *in vitro* experiments. However the exploitation of this attractive feature requires the reduction of the size of this class of materials.

4.3.3.3 Gadolinium metal-organic frameworks (Gd-MOFs)

Metal-organic framework (MOF) designates crystalline hybrid materials built from metal ion connectors and polydentate bridging ligands. One-, two- or three-dimensional nanometric structures can be obtained according to the experimental conditions and to the nature of the metal ions and the organic ligands. These hybrid materials exhibit an interesting potential for a large range of applications (gas purification and separation, catalysis, optical/magnetic sensors) since their behavior can be accurately tuned by a judicious choice of building blocks. When the metal ion is Gd^{3+} or Mn^{2+}, nanoscale MOFs have been proposed as a new class of imaging probes [128–130]. Although the number of water molecules in the inner coordination sphere of each gadolinium ion in the MOF structure, which depends on the nature of the polydendate ligand, is different for $[Gd_2(bhc)(H_2O)_6]$ MOF nanoparticles (bhc for benzene hexacarboxylate) and for $[Gd(H_2cmp)(H_2O)]$ (H_5cmp is (carboxymethyl)iminodi-(methylphosphonic acid)), the longitudinal relaxivity r_1 and the transverse relaxivity r_2 are very similar (r_1 = 1.50 $mM^{-1}s^{-1}$ and r_2 = 122.6 $mM^{-1}s^{-1}$ for $[Gd_2(bhc)(H_2O)_6]$, r_1 = 1.08 $mM^{-1}s^{-1}$ and r_2 = 121.7 $mM^{-1}s^{-1}$ for $[Gd(H_2cmp)(H_2O)]$) [128, 131]. This indicates that the water proton relaxation in these Gd-MOFs occurs through an outer sphere mechanism. Moreover, the r_1 values of these MOFs are very low (probably due to the hindered diffusion of water molecules through the porous MOFs and an inadequate exchange rate between Gd-coordinated water molecules and the bulk water) while high values for r_2 are observed whatever the external magnetic field, in contrast to gadolinium oxide, fluoride, or carbonate nanoparticles (see Sections 4.3.2, 4.3.3.1, and 4.3.3.2). These Gd-MOFs are, therefore, better suited for T_2-weighted MRI.

However, other Gd-MOF nanostructures have a potential for T_1-weighted MRI since higher r_1 values and r_2/r_1 ratio close to 1 are observed for aqueous suspensions of [Gd(1,4-bdc)(H_2O)$_2$] nanorods (1,4-bdc for benzene-1,4-dicarboxylate) and for [Gd(1,2,4-btc)(H_2O)$_3$] nanoplates [129]. As revealed by the inverse size dependence of r_1 and r_2 relaxivities obtained for [Gd(1,4-bdc)(H_2O)$_2$] nanorods, these high r_1 values result mainly from the contribution of Gd^{3+} ions at or near the surface of the MOFs since Gd^{3+} ions inside the material may have a decreased water exchange because of hindered diffusion of water molecules. The magnetic properties of MOFs can, therefore, be controlled by the structure of the MOFs, which can behave as T_1 or T_2 contrast agent for MRI. In addition to this attractive feature, MOFs can be designed as a multifunctional platform combining FI and MRI when their synthesis is performed from a mixture of gadolinium salt and luminescent rare earth salt (Eu^{3+}, Tb^{3+}). Despite the great potential of Gd-MOFs (eventually doped by Eu^{3+} or Tb^{3+}), their use for *in vivo* imaging could be impeded by their large size since in most cases one of the dimensions of Gd-MOF structures is larger than 100 nm.

The functionalization of Gd-MOFs constitutes another crucial issue that should be addressed for *in vivo* application. The research group of Lin succeeded in embedding Mn-MOF in a polysiloxane shell to stabilize them and to facilitate their post-functionalization with an organic fluorophore and a cell-targeting peptide [130]. The functionalization of Gd-MOF nanoparticles ([Gd(1,4-bdc)(H_2O)$_2$]) by well-defined polymers, including poly[*N*-(hydroxypropyl) methacrylamide], poly(*N*-isopropylacrylamide), polystyrene, poly(2-(dimethylamino)ethyl acrylate), poly(((poly)ethylene glycol) methyl ether acrylate), and poly(acrylic acid), was performed by reversible addition-fragmentation chain transfer (RAFT) polymerization, which is one of the most versatile living radical polymerization techniques. Besides the ability to obtain well-defined polymers, RAFT polymerization provides polymer ended by thiolate group after reduction of thiocarbonylthio group in basic conditions. The thiolate groups favor the attachment to the Gd-MOF nanoparticles through vacant orbitals on the Gd^{3+} ions at the surface [132]. The RAFT polymer coated Gd-MOF nanoparticles behave as positive contrast agents. Moreover, the encapsulation of Gd-MOF nanoparticles in hydrophilic RAFT polymer shell (e.g., poly[*N*-(hydroxypropyl)methacrylamide]) induces an increase in the r_1 and

r_2 relaxivities, which is due to the increased water retention in the hydrophilic polymer coating. The r_1 values (r_1 = 105.4 $mM^{-1}s^{-1}$ for Gd-MOF nanoparticles coated with high molecular weight poly[*N*-(hydroxypropyl)methacrylamide]) are significantly higher than both the uncoated Gd-MOF nanoparticles (r_1 = 9.9 $mM^{-1}s^{-1}$ for uncoated Gd-MOF nanoparticles) and clinically-employed contrast agents (Magnevist (r_1 = 13.4 $mM^{-1}s^{-1}$) and Multihance® (r_1 = 19.5 $mM^{-1}s^{-1}$)). RAFT polymerization technique also enables the encapsulation of Gd-MOF nanoparticles (width: 20–25 nm; length: 100–150 nm) in a multifunctional copolymer poly(*N*-isopropylacrylamide)-*co*-poly(*N*-acryloxysuccinimide)-*co*-poly(fluorescein *O*-methacrylate) (thickness: 9 nm) [133].

In addition to the enhancement of the longitudinal relaxivity r_1 (see above), such encapsulation renders Gd-MOF nanoparticles fluorescent (due to the presence of fluorescein) and favors, due to the succinimide functionality, the grafting of both chemotherapeutic (antineoplastic drug (MTX)) and biotargeting groups (GRGDS-NH_2, which targets $\alpha_v\beta_3$ integrins). It was demonstrated that the RAFT copolymer modified Gd-MOF nanoparticles containing the targeting moiety, GRGDS-NH_2, showed active targeting toward FITZ-HSA tumor cells, which overexpress $\alpha_v\beta_3$ integrins. Gd-MOF nanoparticles modified with the RAFT copolymer containing MTX showed dose-dependent treatment of FITZ-HSA cancer cells established by cell viability measurements. Such functionalization strategy renders Gd-MOF nanoparticles very attractive for biomedical applications.

Although they cannot be ranked among MOF nanoparticles, a new family of amorphous, bioinorganic nanoparticles, which are self-assembled from nucleotides and rare earth ions, has been recently described [134]. Due to the large range of properties that exhibit rare earth elements, their combination with nucleotides leads to a wide variety of nanoparticles (size: 40–180 nm depending on the components) with versatile functions as exemplified by sensitized luminescence (when Tb^{3+} ions are used) and enhanced MRI imaging properties (when Gd^{3+} ions are used). The aqueous dispersions of nucleotide/Gd^{3+} nanoparticles showed higher contrast in magnetic resonance images compared to those of Magnevist® per mM of Gd^{3+}. The longitudinal relaxivities (r_1) values of 5'-GMP/Gd^{3+} and 5'-AMP/Gd^{3+} nanoparticles were determined as 13.4 and 12.0 $mM^{-1}s^{-1}$, respectively. These values are twice larger than that of Magnevist® (5.4 $mM^{-1}s^{-1}$ at 0.3 T), and it indicates improved relaxation of water protons in the coordination networks and large rotational correlation time of nanoparticles. Moreover, these nanoparticles display adaptive characteristics and can introduce functional materials such as fluorescent dyes and metal nanoparticles inside. For these reasons, they appear as promising candidates for biomedical applications.

4.4 Conclusion and Future Outlook

Though many outstanding results have been achieved recently, the use of nanoparticles containing rare earth ions as MRI contrast agent is still in its infancy. Many of the studies are currently dealing with *in vitro* analysis or *in vivo* "proof-of-principles." Deeper *in vivo* studies on relevant animal models and clinical trials will certainly give rise to optimization in the design, stability, and contrast enhancement of this new class of contrast agents.

In addition to active and passive targeting for molecular imaging, many nanoparticles containing rare earth ions, combining both imaging and therapeutic capacities, are under development. These multifunctional nanoparticles open new possibilities for the so-called "personalized nanomedicine," where both diagnosis and treatment can be customized for different patients.

As the size range of currently designed nanomaterials is in the window of vascular system (greater than 20 nm), they mainly target vascular-related modifications such as angiogenesis associated with tumor progression or atherosclerosis. With the development of smaller nanoparticles containing rare earth ions (less than 20 nm) that can extravasate from the endothelium to the interstitial space, tumor cell targeting can also be envisioned.

Different strategies are currently under investigation to combine active targeting with therapy: (i) entrapping pharmaceutical compounds within nanomaterials to enable selective drug delivery, (ii) taking advantages of unique physical properties of these nanomaterials to perform selective therapy, or (iii) combining both.

As the use of lanthanide-based nanoparticles is relatively recent, their use in multifunctional nanoplatforms has been rarely investigated so far. Nevertheless, a few recent studies reported are highly promising. For example, a successful attempt to administrate drugs locally has been reported for atherosclerosis treatment in a rabbit model using perfluorocarbon nanoparticles (175–220 nm) loaded with Gd ions as contrast agents and fumagillin as drugs [135].

Other highly promising strategies with minimum side effects rely on the exploitation of unique physical properties of these nanosize carriers for efficient and selective radiotherapy, phototherapy, or neutron therapy. Zielhuis *et al.* have successfully designed liposomes (130 nm) combining paramagnetic gadolinium ions and radionuclides rare earth ions holmium-166 (beta- and gamma-emitter with highly paramagnetic properties) for both imaging and radiotherapy [136].

Several recent studies have also demonstrated the possibility to combine imaging and photothermal therapy, using rare earth-containing nanoparticles with absorbing nanomaterials such as gold [72, 137] or to a lesser extent carbon [138]. The principle relies in the conversion of absorbed light into local heat strong enough to damage nanoparticles-containing cells. This is possible thanks to strong light-absorption properties of nanomaterials (in particular, gold nanomaterials), known as surface plasmon resonance (SPR). These nanomaterials are classically designed to tune the SPR peak in the NIR wavelength region, where the depth of penetration into tissues is higher. The advantages of such technique are minimal invasiveness and increase in selectivity, as damages are only done in the vicinity of the nanomaterials, reducing severe injury of healthy tissues.

Most studies reported so far used relatively large nanoparticles (50 to 450 nm [72, 137, 138]), limiting their use for local injection. There is no doubt, however, that these pioneering works would lead to many innovative nanomaterials more suitable for cell targeting after intravenous injection in clinical applications.

Another attractive feature of gold lies in its high atomic number. As a result, gold element absorbs very strong X-ray photons. Gold nanoparticles appear, therefore, interesting for radiotherapy since their presence in tumor permits the absorption of almost all X-ray photons. Consequently, healthy tissues are more preserved while the tumoricidal effect of X-ray beam is enhanced [66, 67, 70].

Last but not least, neutron therapy appears as an interesting alternative technique yet hardly exploited. The principle relies on the irradiation of compounds with a high neutron capture cross section (neutrophages) by a thermal neutron beam. The irradiation by a thermal neutron beam or the presence of neutrophages compounds is harmless by itself. Selective destruction of surrounding cells can be achieved by combining nanoparticles targeting with thermal neutron beam irradiation. Most therapeutic protocols are nowadays based on molecular boron compounds [139]. However, different groups have reported the possibility to take advantage of γ-rays emission and Auger electrons of gadolinum isotopes (^{155}Gd and ^{157}Gd) upon interaction with a thermal neutron [109, 140–143]. In particular, the *in vitro* study of Tillement's group showed the therapeutic potential of hybrid nanoparticles combining high contrast properties and long blood half-life [104] and a size (<10 nm in diameter) compatible with partial extravasation from the vasculature and cell targeting.

Multifunctional rare earth based nanoparticles are far from clinical use, but different routes to achieve that goal are under investigations by many multidisciplinary joint research programs, starting from chemical design and synthesis to *in vitro* and *in vivo* toxicity and therapeutic efficacy analysis. The aim is to be able to get diagnosis as early as possible and to treat the pathology with high efficacy and limited side effects. Rare earth based multifunctionnal nanoplatforms will open a possible route to achieve high contrast properties (and thus lowering the detection sensitivity threshold), as well as high specificity toward the targeted pathological sites. The near future will certainly lead to the design of more innovative rare earth based multifunctionnal nanoplatforms for early diagnosis and image-guided therapy [144, 145], most presumably in oncology and atherosclerosis diseases.

References

1. Brigger, I.; Dubernet, C.; Couvreur, P. (2002), *Adv. Drug Deliv. Rev.*, **54**, 631–651.
2. Sunderland, C. J.; Steiert, M.; Talmadge, E. J.; Derfus, M. A.; Barry, S. E. (2006) *Drug Dev. Res.*, **67**, 70–93.
3. Whitesides, G. M. (2003) *Nat. Biotechnol.*, **21**, 1161–1165.
4. Ferrari, M. (2005) *Nat. Rev.*, **5**, 161–171.
5. Medintz, I. L.; Uyeda, H. T.; Goldman, E. R.; Mattoussi, H. (2005) *Nat. Mater.*, **4**, 435–446.
6. Lewin, M.; Carlesso, N.; Tung, C. H.; Tang, X. W.; Cory, D.; Scadden, D. T.; *et al.* (2000) *Nat. Biotechnol.*, **18**, 410–414.
7. Kircher, M. F.; Mahmood, U.; King, R. S.; Weissleder, R.; Josephson, L. (2003) *Cancer Res.*, **63**, 8122–8125.

8. Caravan, P. (2006) *Chem. Soc. Rev.*, **35**, 512–523.
9. Caravan, P.; Ellison, J. J.; McMurry, T. J.; Lauffer, R. B. (1999) *Chem. Rev.*, **99**, 2293–2352.
10. Bottrill, M.; Kwok, L.; Long, N. J. (2006) *Chem. Soc. Rev.*, **35**, 557–571.
11. Nahrendorf, M.; Zhang, H.; Hembrador, S.; Panizzi, P.; Sosnovik, D. E.; Aikawa, E.; *et al.* (2008) *Circulation*, **117**, 379–387.
12. Weissleder, R. (2006) *Science*, **312**, 1168–1171.
13. Cai, W.; Chen, X. (2007) *Small*, **3**, 1840–1854.
14. Riehemann, K.; Schneider, S. W.; Luger, T. A.; Godin, B.; Ferrari, M.; Fuchs, H. (2009) *Angew. Chem. Int. Ed.*, **48**, 872–897.
15. Stöber, W.; Fink, A.; Bohn, E. (1968) *J. Colloid. Interface Sci.*, **26**, 62–69.
16. Bagwe, R. P.; Hilliard, L. R.; Tan, W. (2006) *Langmuir*, **22**, 4357–4362.
17. van Blaaderen, A.; Vrij, A. (1992) *Langmuir*, **8**, 2921–2931.
18. Ow, H.; Larson, D. R.; Srivastava, M.; Baird, B. A.; Webb, W. W.; Wiesner, U. (2005) *Nano Lett.*, **5**, 113–117.
19. Burns, A. A.; Vider, J.; Ow, H.; Herz, E.; Penate–Medina, O.; Baumgart, M.; Larson, S. M.; *et al.* (2009) *Nano Lett.*, **9**, 442–448.
20. Zhao, X.; Tapec–Dytioco, R.; Tan, W. (2003) *J. Am. Chem. Soc.*, **125**, 11474–11475.
21. Wang, L.; Tan, W. (2006) *Nano Lett.*, **6**, 84–88.
22. Santra, S.; Bagwe, R. P.; Dutta, D.; Stanley, J. T.; Walter, G. A.; Tan, W.; *et al.* (2005) *Adv. Mater.*, **17**, 2165–2169.
23. Rieter, W. J.; Kim, J. S.; Taylor, K. M.; An, H.; Lin, W.; Tarrant, T. (2007) *Angew. Chem. Int. Ed.*, **46**, 3680–3682.
24. Kim, J. S.; Rieter, W. J.; Taylor, K. M.; An, H.; Lin, W. (2007) *J. Am. Chem. Soc.*, **129**, 8962–8963.
25. Delgado–Pinar, E.; Frias, J. C.; Jimenez-Borreguero, L. J.; Albelda, M. T.; Alarcon, J.; Garcia-Espana, E. (2007) *Chem. Commun.*, 3392–3394.
26. Voisin, P.; Ribot, E. J.; Miraux, S.; Bouzier-Sore, A.-K.; Lahitte, J.-F.; Bouchaud, V.; *et al.* (2007) *Bioconjug. Chem.*, **18**, 1053–1063.
27. Endres, P. J.; Paunesku, T.; Vogt, S.; Meade, T. J.; Woloschak, G. E. (2007) *J. Am. Chem. Soc.*, **129**, 15760–15761.
28. Taylor, K. M.; Kim, J. S.; Rieter, W. J.; An, H.; Lin, W. (2008) *J. Am. Chem. Soc.*, **130**, 2154–2155.
29. Chen, L. X.; Liu, T.; Thurnauer, M. C.; Csencsits, R.; Rajh, T. (2002) *J. Phys. Chem. B*, **106**, 8539–8546.

30. Bae, K. H.; Kim, Y. B.; Lee, Y.; Hwang, J. Y.; Park, H. W.; Park, T. G. (2010) *Bioconjug. Chem.*, **21**, 505–512.

31. Rehor, I.; Kubicek, V.; Kotek, J.; Hermann, P.; Lukes, I.; Szakova, J.; *et al.* (2009) *J. Mater. Chem.*, **19**, 1494–1500.

32. Huang, D. M.; Hung, Y.; Ko, B. S.; Hsu, S. C.; Chen, W. H.; Chien, C. L.; *et al.* (2005) *FASEB J.*, **19**, 2014–2016.

33. Chung, T. H.; Wu, S. H.; Yao, M.; Lu, C. W.; Lin, Y. S.; Hung, Y.; *et al.* (2007) *Biomaterials*, **28**, 2959–2966.

34. Hsiao, J. K.; Tsai, C. P.; Chung, T. H.; Hung, Y.; Yao, M.; Liu, H. M.; *et al.* (2008) *Small*, **4**, 1445–1452.

35. Carniato, F.; Tei, L.; Dastru, W.; Marchese, L.; Botta, M. (2009) *Chem. Commun.*, 1246–1248.

36. Tsai, C. P.; Hung, Y.; Chou, Y. H.; Huang, D. M.; Hsiao, J. K.; Chang, C.; *et al.* (2008) *Small*, **4**, 186–191.

37. Carniato, F.; Tei, L.; Dastru, W.; Marchese, L.; Botta, M. (2009) *Chem. Commun.*, 1246–1248.

38. Lin, Y. S.; Hung, Y.; Su, J. K.; Lee, R.; Chang, C.; Lin, M. L.; *et al.* (2004) *J. Phys. Chem. B*, **108**, 15608–15611.

39. Wu, C.; Hong, J.; Guo, X.; Huang, C.; Lai, J.; Zheng, J.; *et al.* (2008) *Chem. Commun.*, 750–752.

40. Lu, F.; Wu, S. H.; Hung, Y.; Mou, C. Y. (2009) *Small*, **5**, 1408–1413.

41. Daldrup-Link, H. E.; Rudelius, M.; Oostendorp, R. A.; Settles, M.; Piontek, G.; Metz, S.; *et al.* (2003) *Radiology*, **228**, 760–767.

42. Crich, S. G.; Biancone, L.; Cantaluppi, V.; Duo, D.; Esposito, G.; Russo, S.; *et al.* (2004) *Magn. Reson. Med.*, **51**, 938–944.

43. Bruchez, M., Jr.; Moronne, M.; Gin, P.; Weiss, S.; Alivisatos, A. P. (1998) *Science*, **281**, 2013–2016.

44. Chan, W. C.; Nie, S. (1998) *Science*, **281**, 2016–2018.

45. Maxwell, D. J.; Taylor, J. R.; Nie, S. (2002) *J. Am. Chem. Soc.*, **124**, 9606–9612.

46. Derfus, A. M.; Chan, W. C. W.; Bhatia, S. N. (2004) *Nano Lett.*, **4**, 11–18.

47. Lewinski, N.; Colvin, V.; Drezek, R. (2008) *Small*, **4**, 26–49.

48. Kobayashi, H.; Hama, Y.; Koyama, Y.; Barrett, T.; Regino, C. A.; Urano, Y.; *et al.* (2007) *Nano Lett.*, **7**, 1711–1716.

49. Choi, H. S.; Liu, W.; Misra, P.; Tanaka, E.; Zimmer, J. P.; Itty Ipe, B.; *et al.* (2007) *Nat. Biotechnol.*, **25**, 1165–1170.

50. Roullier, V.; Clarke, S.; You, C.; Pinaud, F.; Gouzer, G. G.; Schaible, D.; *et al.* (2009) *Nano Lett.*, **9**, 1228–1234.
51. Jin, T.; Yoshioka, Y.; Fujii, F.; Komai, Y.; Seki, J.; Seiyama, A. (2008) *Chem. Commun.*, 5764–5766.
52. Gerion, D.; Herberg, J.; Bok, R.; Gjersing, E.; Ramon, E.; Maxwell, R.; *et al.* (2007) *J. Phys. Chem. C*, **111**, 12542–12551.
53. Bakalova, R.; Zhelev, Z.; Aoki, I.; Masamoto, K.; Mileva, M.; Obata, T.; *et al.* (2008) *Bioconjug. Chem.*, **19**, 1135–1142.
54. Koole, R.; van Schooneveld, M. M.; Hilhorst, J.; Castermans, K.; Cormode, D. P.; Strijkers, G. *et al.* (2008) *Bioconjug. Chem.*, **19**, 2471–2479.
55. Mulder, W. J.; Koole, R.; Brandwijk, R. J.; Storm, G.; Chin, P. T.; Strijkers, G. J.; *et al.* (2006) *Nano Lett.*, **6**, 1–6.
56. van Tilborg, G. A.; Mulder, W. J.; Chin, P. T.; Storm, G.; Reutelingsperger, C. P.; Nicolay, K.; *et al.* (2006) *Bioconjug. Chem.*, **17**, 865–868.
57. Mulder, W. J.; Strijkers, G. J.; van Tilborg, G. A.; Cormode, D. P.; Fayad, Z. A.; Nicolay, K. (2009) *Acc. Chem. Res.*, **42**, 904–914.
58. van Schooneveld, M. M.; Vucic, E.; Koole, R.; Zhou, Y.; Stocks, J.; Cormode, D. P.; *et al.* (2008) *Nano Lett.*, 8, 2517–2525.
59. Cormode, D. P.; Skajaa, T.; van Schooneveld, M. M.; Koole, R.; Jarzyna, P.; Lobatto, M. E.; *et al.* (2008) *Nano Lett.*, **8**, 3715–3723.
60. Frias, J. C.; Williams, K. J.; Fisher, E. A.; Fayad, Z. A. (2004) *J. Am. Chem. Soc.*, **126**, 16316–16317.
61. Frias, J. C.; Ma, Y.; Williams, K. J.; Fayad, Z. A.; Fisher, E. A. (2006) *Nano Lett.*, **6**, 2220–2224.
62. Boisselier, E.; Astruc, D. (2009) *Chem. Soc. Rev.*, **38**, 1759–1782.
63. Ghosh, S. K.; Pal, T. (2007) *Chem. Rev.*, **107**, 4797–4862.
64. Lal, S.; Clare, S. E.; Halas, N. J. (2008) *Acc. Chem. Res.*, **41**, 1842–1851.
65. Jain, P. K.; El-Sayed, I. H.; El-Sayed, M. A. (2007) *Nanotoday*, **2**, 18–29.
66. Hainfeld, J. F.; Slatkin, D. N.; Focella, T. M.; Smilowitz, H. M. (2006) *Br. J. Radiol.*, **79**, 248–253.
67. Hainfeld, J. F.; Slatkin, D. N.; Smilowitz, H. M. (2004) *Phys. Med. Biol.*, **49**, N309–315.
68. Debouttière, P. J.; Roux, S.; Vocanson, F.; Billotey, C.; Beuf, O.; Favre-Réguillon, A.; *et al.* (2006) *Adv. Funct. Mater.*, **16**, 2330–2339.
69. Alric, C.; Taleb, J.; Le Duc, G.; Mandon, C.; Billotey, C.; Le Meur-Herland, A.; *et al.* (2008) *J. Am. Chem. Soc.*, **130**, 5908–5915.

70. Alric, C.; Serduc, R.; Mandon, C.; Taleb, J.; Le Duc, G.; Le Meur-Herland, A.; *et al.* (2008) *Gold Bull.*, **41**, 90–97.

71. Massue, J.; Quinn, S. J.; Gunnlaugsson, T. (2008) *J. Am. Chem. Soc.*, **130**, 6900–6901.

72. Lim, Y. T.; Cho, M. Y.; Choi, B. S.; Lee, J. M.; Chung, B. H. (2008) *Chem. Commun.*, 4930–4932.

73. Huber, M. M.; Staubli, A. B.; Kustedjo, K.; Gray, M. H.; Shih, J.; Fraser, S. E.; *et al.* (1998) *Bioconjug. Chem.*, **9**, 242–249.

74. Yan, G. P.; Liu, M. L.; Li, L. Y. (2005) *Bioconjug. Chem.*, **16**, 967–971.

75. Bull, S. R.; Guler, M. O.; Bras, R. E.; Venkatasubramanian, P. N.; Stupp, S. I.; Meade, T. J. (2005) *Bioconjug. Chem.*, **16**, 1343–1348.

76. Bull, S. R.; Guler, M. O.; Bras, R. E.; Meade, T. J.; Stupp, S. I. (2005) *Nano Lett.*, **5**, 1–4.

77. Anderson, E. A.; Isaacman, S.; Peabody, D. S.; Wang, E. Y.; Canary, J. W.; Kirshenbaum, K. (2006) *Nano Lett.*, **6**, 1160–1164.

78. Langereis, S.; de Lussanet, Q. G.; van Genderen, M. H. P.; Backes, W. H.; Meijer, E. W. (2004) *Macromolecules*, **37**, 3084–3091.

79. Talanov, V. S.; Regino, C. A.; Kobayashi, H.; Bernardo, M.; Choyke, P. L.; Brechbiel, M. W. (2006) *Nano Lett.*, **6**, 1459–1463.

80. Mulder, W. J.; Strijkers, G. J.; Griffioen, A. W.; van Bloois, L.; Molema, G.; Storm, G.; *et al.* (2004) *Bioconjug. Chem.*, **15**, 799–806.

81. Accardo, A.; Tesauro, D.; Roscigno, P.; Gianolio, E.; Paduano, L.; D'Errico, G.; *et al.* (2004) *J. Am. Chem. Soc.*, **126**, 3097–3107.

82. Parac-Vogt, T. N.; Kimpe, K.; Laurent, S.; Piérart, C.; Vander Elst, L.; Muller, R. N.; *et al.* (2004) *Eur. J. Inorg. Chem.*, **17**, 3538–3543.

83. Reynolds, C. H.; Annan, N.; Beshah, K.; Huber, J. H.; Shaber, S. H.; Lenkinski, R. E.; *et al.* (2000) *J. Am. Chem. Soc.*, **122**, 8940–8945.

84. Turner, J. L.; Pan, D.; Plummer, R.; Chen, Z.; Whittaker, A. K.; Wooley, K. L. (2005) *Adv. Funct. Mater.*, **15**, 1248–1254.

85. Platas-Iglesias, C.; Vander Elst, L.; Zhou, W.; Muller, R. N.; Geraldes, C. F.; Maschmeyer, T.; *et al.* (2002) *Chemistry*, **8**, 5121–5131.

86. Bolskar, R. D.; Benedetto, A. F.; Husebo, L. O.; Price, R. E.; Jackson, E. F.; Wallace, S.; *et al.* (2003) *J. Am. Chem. Soc.*, **125**, 5471–5478.

87. Mikawa, M.; Kato, H.; Okumura, M.; Narazaki, M.; Kanazawa, Y.; Miwa, N.; *et al.* (2001) *Bioconjug. Chem.*, **12**, 510–514.

88. Toth, E.; Bolskar, R. D.; Borel, A.; Gonzalez, G.; Helm, L.; Merbach, A. E.; *et al.* (2005) *J. Am. Chem. Soc.*, **127**, 799–805.

89. Sitharaman, B.; Bolskar, R. D.; Rusakova, I.; Wilson, L. J. (2004) *Nano Lett.*, **4**, 2373–2378.

90. Sitharaman, B.; Kissell, K. R.; Hartman, K. B.; Tran, L. A.; Baikalov, A.; Rusakova, I.; *et al.* (2005) *Chem. Commun.*, 3915–3917.

91. Hartman, K. B.; Laus, S.; Bolskar, R. D.; Muthupillai, R.; Helm, L.; Toth, E.; *et al.* (2008) *Nano Lett.*, **8**, 415–419.

92. Balkus, K. J.; Shi, J. (1996) *Langmuir*, **12**, 6277–6281.

93. Vuu, K.; Xie, J.; McDonald, M. A.; Bernardo, M.; Hunter, F.; Zhang, Y.; *et al.* (2005) *Bioconjug. Chem.*, **16**, 995–999.

94. Engström, M.; Klasson, A.; Pedersen, H.; Vahlberg, C.; Käll, P. O.; Uvdal, K. (2006) *Magn. Reson. Mater. Phys.*, **19**, 180–186.

95. McDonald, M. A.; Watkin, K. L. (2003) *Invest. Radiol.*, **38**, 305–310.

96. Roberts, D.; Zhu, W. L.; Frommen, C. M.; Rosenzweig, Z. (2000) *J. Appl. Phys.*, **87**, 6208–6210.

97. Evanics, F.; Diamente, P. R.; van Veggel, F.; Stanisz, G. J.; Prosser, R. S. (2006) *Chem. Mater.*, **18**, 2499–2505.

98. Hu, K. W.; Jhang, F.; Su, C. H.; Yeh, C. S. (2009) *J. Mater. Chem.*, **19**, 2147–2153.

99. Li, I. F.; Su, C. H.; Sheu, H. S.; Chiu, H. C.; Lo, Y. W.; Lin, W. T.; *et al.* (2008) *Adv. Funct. Mater.*, **18**, 766–776.

100. Bazzi, R.; Flores, M. A.; Louis, C.; Lebbou, K.; Zhang, W.; Dujardin, C.; *et al.* (2004) *J. Colloid. Interface Sci.*, **273**, 191–197.

101. Bazzi, R.; Flores-Gonzalez, M. A.; Louis, C.; Lebbou, K.; Dujardin, C.; Brenier, A.; *et al.* (2003) *J. Lumin.*, **102**, 445–450.

102. Louis, C.; Bazzi, R.; Marquette, C. A.; Bridot, J.-L.; Roux, S.; Ledoux, G.; *et al.* (2005) *Chem. Mater.*, **17**, 1673–1682.

103. Louis, C.; Roux, S.; Ledoux, G.; Dujardin, C.; Tillement, O.; Cheng, B. L.; *et al.* (2006) *Chem. Phys. Lett.*, **429**, 157–160.

104. Bridot, J. L.; Faure, A. C.; Laurent, S.; Riviere, C.; Billotey, C.; Hiba, B.; *et al.* (2007) *J. Am. Chem. Soc.*, **129**, 5076–5084.

105. Faure, A. C.; Dufort, S.; Josserand, V.; Perriat, P.; Coll, J. L.; Roux, S.; *et al.* (2009) *Small*, **5**, 2565–2575.

106. Faure, A. C.; Hoffmann, C.; Bazzi, R.; Goubard, F.; Pauthe, E.; Marquette, C. A.; *et al.* (2008) *ACS Nano*, **2**, 2273–2282.

107. Le Duc, G.; Roux, S.; Tillement, O.; Perriat, P.; Billotey, C.; Mandon, C.; *et al.* In *Brevet français*; 07.58348: France, 16/10/2007.

108. Roux, S.; Tillement, O.; Billotey, C.; Coll, J. L.; Le Duc, G.; Marquette, C. A.; *et al.* (2010) *Int. J. Nanotechnol.*, **7**, 781–801.

109. Bridot, J.-L.; Dayde, D.; Riviere, C.; Mandon, C.; Billotey, C.; Lerondel, S.; *et al.* (2009) *J. Mater. Chem.*, **19**, 2328–2335.

110. Soderlind, F.; Pedersen, H.; Petoral, R. M., Jr.; Kall, P. O.; Uvdal, K. (2005) *J. Colloid. Interface Sci.*, **288**, 140–148.

111. Pedersen, H.; Soderlind, F.; Petoral, R. M.; Uvdal, K.; Kall, P. O.; Ojamae, L. (2005) *Surf. Sci.*, **592**, 124–140.

112. Hermanson, G. T. (2008) *Bioconjugate Techniques*, 2nd ed., Academic Press, Amsterdam.

113. Fortin, M. A.; Petoral, R. M.; Soderlind, F.; Klasson, A.; Engstrom, M.; Veres, T.; *et al.* (2007) *Nanotechnology*, **18**, 395501.

114. Petoral, R. M.; Soderlind, F.; Klasson, A.; Suska, A.; Fortin, M. A.; Abrikossova, N.; *et al.* (2009) *J. Phys. Chem. C*, **113**, 6913–6920.

115. Fizet, J.; Riviere, C.; Bridot, J. L.; Charvet, N.; Louis, C.; Billotey, C.; *et al.* (2009) *J. Nanosci. Nanotechnol.*, **9**, 1–9.

116. Norek, M.; Kampert, E.; Zeitler, U.; Peters, J. A. (2008) *J. Am. Chem. Soc.*, **130**, 5335–5340.

117. Norek, M.; Pereira, G. A.; Geraldes, C. F. G. C.; Denkova, A.; Zhou, W.; Peters, J. A. (2007) *J. Phys. Chem. C*, **111**, 10240–10246.

118. Laurent, S.; Forge, D.; Port, M.; Roch, A.; Robic, C.; Vander Elst, L.; *et al.* (2008) *Chem. Rev.*, **108**, 2064–2110.

119. Miyawaki, J.; Yudasaka, M.; Imai, H.; Yorimitsu, H.; Isobe, H.; Nakamura, E.; *et al.* (2006) *J. Phys. Chem. B*, **110**, 5179–5181.

120. Sánchez, P.; Valero, E.; Gálvez, N.; Domínguez-Vera, J. M.; Marinone, M.; Poletti, G.; *et al.* (2009) *Dalton Trans.*, 800–804.

121. Huang, C. C.; Liu, T. Y.; Su, C. H.; Lo, Y. W.; Chen, J. H.; Yeh, C. S. (2008) *Chem. Mater.*, **20**, 3840–3848.

122. Diamente, P. R.; Burke, R. D.; van Veggel, F. C. (2006) *Langmuir*, **22**, 1782–1788.

123. Chatterjee, D. K.; Rufaihah, A. J.; Zhang, Y. (2008) *Biomaterials*, **29**, 937–943.

124. Nyk, M.; Kumar, R.; Ohulchanskyy, T. Y.; Bergey, E. J.; Prasad, P. N. (2008) *Nano Lett.*, **8**, 3834–3838.

125. Chen, Z. G.; Tian, Q. W.; Cao, T. Y.; Xu, C. J.; Li, F. Y. (2009) *Anal. Chem.*, **81**, 8687–8694.

126. Kumar, R.; Nyk, M.; Ohulchanskyy, T. Y.; Flask, C. A.; Prasad, P. N. (2009) *Adv. Funct. Mater.*, **19**, 853–859.

127. Zhou, J.; Sun, Y.; Du, X.; Xiong, L.; Hu, H.; Li, F. (2010) *Biomaterials*, **31**, 3287–3295.

128. Taylor, K. M. L.; Jin, A.; Lin, W. (2008) *Angew. Chem. Int. Ed.*, **47**, 7722–7725.

129. Rieter, W. J.; Taylor, K. M. L.; An, H.; Lin, W. (2006) *J. Am. Chem. Soc.*, **128**, 9024–9025.

130. Taylor, K. M. L.; Rieter, W.J.; Lin, W. (2008) *J. Am. Chem. Soc.*, **130**, 14358–14359.

131. Pereira, G. A.; Peters, J. A.; Almeida Paz, F. A.; Rocha, J.; Geraldes, F. G. C. (2010) *Inorg. Chem.*, **49**, 2969–2974.

132. Rowe, M. D.; Chang, C.-C.; Thamm, D. H.; Kraft, S. L.; Harmon, J. F.; Vogt, A. P.; *et al.* (2009) *Langmuir*, **26**, 9487–9499.

133. Rowe, M. D.; Thamm, D. H.; Kraft, S. L.; Boyes, S. G. (2009) *Biomacromolecules*, **10**, 983–993.

134. Nishiyabu, R.; Hashimoto, N.; Cho, T.; Watanabe, K.; Yasugana, T.; Endo, A.; *et al.* (2009) *J. Am. Chem. Soc.*, **131**, 2151–2158.

135. Winter, P. M.; Neubauer, A. M.; Caruthers, S. D.; Harris, T. D.; Robertson, J. D.; Williams, T. A.; *et al.* (2006) *Arterioscler. Thromb. Vasc. Biol.*, **26**, 2103–2109.

136. Zielhuis, S. W.; Seppenwoolde, J. H.; Mateus, V. A.; Bakker, C. J.; Krijger, G. C.; Storm, G.; *et al.* (2006) *Cancer Biother. Radiopharm.*, **21**, 520–527.

137. Hu, K.-W.; Jhang, F.-Y.; Su, C.-H.; Yeh, C.-S. (2009) *J. Mater. Chem.*, **19**, 2147–2153.

138. Huang, C. C.; Su, C. H.; Li, W. M.; Liu, T. Y.; Chen, J. H.; Yeh, C. S. (2009) *Adv. Funct. Mater.*, **19**, 249–258.

139. Barth, R. F.; Coderre, J. A.; Vicente, M. G. H.; Blue, T. E. (2005) *Clin. Cancer Res.*, **11**, 3987–4002.

140. Le, U. M.; Cui, Z. (2006) *Int. J. Pharm.*, **320**, 96–103.

141. Enger, S. A.; Rezaei, A.; Munck af Rosenschöld, P.; Lundqvist, H. (2006) *Med. Phys.*, **33**, 46–51.

142. Matsumura, A.; Zhang, T.; Nakai, K.; Endo, K.; Kumada, H.; Yamamoto, T.; *et al.* (2005) *J. Exp. Clin. Cancer Res.*, **24**, 93–98.

143. Wangerin, K.; Culbertson, C. N.; Jevremovic, T. (2005) *Health Phys.*, **89**, 135–144.

144. Lux, F.; Mignot, A.; Mowat, P.; Louis, C.; Dufort, S.; Bernhard, C.; *et al.* (2011) *Angew. Chem. Int. Ed.*, **50**, 12299–12303.

145. Le Duc, G.; Miladi, I.; Alric, C.; Mowat, P.; Bräuer-Krisch, E.; Bouchet, A.; *et al.* (2011) *ACS Nano*, doi: 10.1021/nn202797h.

Chapter 5

Rare Earth Nanomaterials in Integrated Modalities Imaging

Y. Zhang and Timothy T.Y. Tan
School of Chemical and Biomedical Engineering, Nanyang Technological University, 62 Nanyang Drive, 637459, Singapore
tytan@ntu.edu.sg

5.1 Introduction and Overview

Molecular imaging is the *in vivo* visualization, characterization, and measurement of biological processes at molecular and cellular levels in humans and other living systems [1–4]. It offers great potential for detecting molecular and cellular changes caused by diseases before the abnormalities are large enough to cause change in the anatomical structure [2]. Moreover, it enables examination and understanding of the integrative functions of molecules, cells, and organ systems and is an efficient tool to monitor the effect of therapeutic treatments [5]. For these merits, much research efforts have been dedicated to this area in recent decades and significant advances have been made.

Rare Earth Nanotechnology
Edited by Timothy T.Y. Tan

ISBN 978-981-4316-30-9 (Hardcover), 978-981-4364-20-1 (eBook)
www.panstanford.com

Molecular imaging modalities such as positron emission tomography (PET), single photon emission computed tomography (SPECT), magnetic resonance imaging (MRI), computed tomography (CT), ultrasound and optical imaging are widely used clinically to understand and diagnose various diseases. The characteristics of every modality are summarized in Table 5.1. Every imaging modality has its own advantages and disadvantages and no single technique possesses full capabilities to obtain comprehensive and accurate biological information. For example, MRI and CT have the advantages of high spatial resolution, but they are limited by low sensitivity. On the contrary, optical imaging has very high sensitivity but suffers from low tissue penetration depths. Therefore, by integrating different imaging modalities into a single system, multimodal imaging method can offset the limitations of single imaging modalities. PET/CT is one of the multimodal imaging instruments. The first PET/CT scanner, developed in 1998 by Townsend and his colleagues in collaboration with Siemens Medical, was commercially available in 2001 [6, 7]. By 2003, PET/CT instruments were available from all major clinical instrument manufactures such as GE, Philips, CTI, and Siemens. Today, more than 95% of new PET scanners installed are integrated PET/CT scanners. SPECT/CT was introduced commercially in 2004 [8]. PET/MRI instruments, which have brought much hope for improved patient safety and imaging capacity over PET/CT, are also on the horizon [9].

Imaging probes are used to monitor and trace molecular processes. With the advent of multimodal technology, design and development of new functional molecular imaging probes are of great importance. There are two prerequisites for functional molecular imaging probes: first, they have to provide high spatial resolution as well as high sensitivity in imaging; second, target-specific molecular probes have to be available [10]. Many possible combinations of imaging modalities result in different types of imaging contrast agents. Increasing number of works have reported the fabrication of different inorganic multimodal contrast agents. Some examples include iron oxide nanoparticles conjugated with dye molecules [11, 12], iron oxide/quantum dots (QDs) heterodimers [13–16], and paramagnetic ion (Mn^{2+})-doped QDs [17, 18].

Rare earth ions possess unique optical and magnetic properties due to their 4f electronic configuration. The $4f^n$ electronic configuration is shielded by an external $5s^25p^6$ subshell, protecting the 4f electronic transitions from external influences. The technological importance of these elements has attracted a lot of interest in magnetic [19] and optical materials [20], as well as biorelated materials [21]. In this chapter, we will first briefly discuss the imaging modality relevant to this chapter. Rare earth based nanomaterials as potential multimodal contrast agents and their applications in these various imaging modalities will be reviewed.

Table 5.1 Characteristics of every modality

Modality	Resolution	Depth	Cost	Time	Imaging agents
PET	1–2 mm	No limit	High	Minutes	Radioisotope (^{18}F, ^{64}Cu, ^{99m}Tc, ^{124}I)
SPECT	1–2 mm	No limit	High	Minutes	Radioisotope (^{18}F, ^{124}I, ^{64}Cu, ^{99m}Tc)
MRI	10–100 μm	No limit	High	Minutes-Hours	Paramagnetic ions (Gd^{3+}, Mn^{2+}); Paramagnetic nanoparticles; Superparamagnetic nanoparticles (iron oxide)
Optical	1 μm	< 400 μm	Low	Seconds-Minutes	Organic dyes; Fluorescent protein; Rare earth chelate; QDs; Rare earth nanomaterials; Carbon nanotube;
CT	50 μm	No limit	Medium	Minutes	Iodine; Gadolinium; Gold nanoparticles; Bismuth sulfide nanoplate;
Ultra-sound	50 μm	Milli-meters	Medium	Minutes	Microbubble; Perfluorocarbon nanoparticles;

Note: Adapted from [22].

5.1.1 Positron Emission Tomography and Single Photon Emission Computed Tomography

PET and SPECT are part of nuclear imaging techniques. They have the advantages of high intrinsic sensitivity and unlimited penetration depth [23]. PET has the additional advantages of being fully quantitative, providing higher spatial resolution than SPECT [24, 25]. Hundreds of radiotracers based on a wide variety of radionuclides due to β- or γ-rays have been developed and tested in animal and clinical studies. Some examples of the radionuclides are summarized in Table 5.2 [26–28]. The half-lives of these radiotracers are short; thus they need to be produced on site, which increases the cost of imaging. For clinical use, these radiotracers are injected into a patient's bloodstream. The amount of the radiotracers injected is so small (ng or μg) that the toxicity is usually not a major issue.

In PET, the tracers are incorporated into a biologically active molecule, such as fluorodeoxyglucose (FDG). The isotope accumulates in an area of interest in the body, and the radiotracers decay by positron emission. After some time, the distribution of the positron-emitting tracer is calculated by tomographic reconstruction procedures. PET can be used to study neuroreceptors in the brain and other body tissues. Clinical studies include tumors of the brain, breast, lung, lower gastrointestinal tract, study of Alzheimer's disease, Parkinson's disease, epilepsy, and coronary artery disease affecting heart muscle metabolism [24, 25, 29].

Table 5.2 Examples of radionuclides

Radionuclide	Half-life time	Application
^{11}C	20.3 min	Used for metabolism studies
^{13}N	9.9 min	Used as 13NH for hear perfusion study
^{15}O	123 s	Used for blood flow
^{18}F	110 min	Quantified tumor cell proliferation, identification of tumor hypoxia
^{52}Fe	8.2 h	Used for PET bone marrow imaging
^{64}Cu	12.8 h	Used for cancer and metabolic disorder
^{99m}Tc	6 h	The most widely used radiopharmaceutical in nuclear medicine
^{122}I	3.76 min	Used for blood flow study
^{131}I	8.1 days	Diagnose thyroid disorders, including cancer

Note: Adapted from [26–28].

SPECT is similar to PET in terms of using radiotracer materials. However, SPECT is based on the detection of γ-emitting isotopes directly, such as ^{99m}Tc, ^{111}In, ^{123}I, ^{125}I, and ^{131}I, whereas PET tracer emits positrons [30]. SPECT costs less than PET scan due to longer half-lives and more easily obtained radiotracers. However, SPECT generates images of lower resolution [24, 30, 31].

5.1.2 Magnetic Resonance Imaging

MRI is popular for diagnostic and drug development due to its greater soft tissue contrast, higher spatial resolution (as good as tens of micrometers for preclinical studies and ~1 mm for clinical studies) and good penetration depth [22, 32]. It does not involve the use of ionizing radiation, which is used in X-ray and CT scan. However, low sensitivity for the detection of targeted agents hampers its usage as compared to PET and SPECT. Owing to the recent development of powerful, innovative MRI contrast agents, it is now more popular as a noninvasive imaging modality.

MRI uses a powerful magnetic field to align the nuclear magnetization of hydrogen atoms in water protons. When the external magnetic field is applied to these protons, which are oriented in random directions, they become aligned with the external magnetic field. They can either align along or opposite to the magnetic field. The oriented spins remain constant and form net magnetization. The magnetization can be disturbed from its equilibrium state using a radio frequency pulse. When the pulse is removed from the system, the protons realign with the equilibrium states, and MRI machines measure these relaxation time [33]. There are two types of relaxation time depending on different pulses used, namely, T_1 (longitudinal relaxation, the time taken for the protons to realign with the external magnetic field) and T_2 (transverse relaxation, the time taken for the protons to exchange energy with other nuclei).

Based on the relaxation times, contrast agents can be categorized into T_1 and T_2 contrast agents. T_1 contrast agents usually include gadolinium-based chelates or nanoparticles [34] and manganese-based chelates or nanoparticles [35]. These contrast agents have unpaired electrons, which give them paramagnetic properties. T_2 contrast agents include superparamagnetic iron oxide nanoparticles and rare earth ions such as Dy(III) [21, 36–39].

Several factors affect the quality of MRI contrast agents: water exchange rate (also called inner-sphere effects), rotational correlation time (how quickly the contrast agent is tumbling in solution), and susceptibility effect (also caller outer-sphere effect) [40].

5.1.3 Optical Imaging

Optical imaging, which is based on the detection of light photons after their interaction with biological tissue, has been used in biomedical diagnostics for years [1, 41]. It provides high sensitivity for detection and is a nonionizing method. An optical imaging instrument is relatively cheap and easy to use. Limitations of optical imaging include low penetration depth and scattering of light in the tissue [42]. Therefore, as the sensitivity of detection is primarily dependent on the fluorphores, the development of novel fluorphores with suitable functionalities and properties is paramount. Ideal fluorphores should have good solubility in biological media, specific association with target molecules or cells, typical absorption wavelengths and fluorescence emissions, high fluorescence quantum yields and lifetimes, and photochemically stable [43] .

A wide variety of fluorophores for bioimaging applications are currently available. Traditional fluorophores include organic dyes, fluorescent proteins, and lanthanide chelates, which are still the most commonly used fluorescent labels. They have small size, good water solubility, and are easy to use [44]. However, these conventional fluorophors suffer from several limitations such as broad absorption spectrum, high photobleaching susceptibility, and poor photochemical stability. Novel inorganic fluorophores, on the contrary, have the potential to overcome these problems. Examples of inorganic fluorophores include QDs, rare earth doped inorganic nanoparticles, and silica nanobeads [45–48].

5.1.4 X-Ray Computed Tomography

X-ray computed tomography is another clinical imaging modality. It employs X-rays, together with computer software, to create images of tissues. The X-ray tube rotates around the object, and the transmission profiles are sampled from up to 360° angles in a digitized form using a detector during imaging acquisition. Tomography was created by computer processing. It was invented by Sir Godfrey Hounsfield, who later received the Nobel Prize in medicine in 1979 [6]. Since then, the modality has gained popularity in medical imaging to supplement X-rays and medical ultrasonography. Contrast agents are used to obtain superior image quality [49]. The disadvantage of X-ray CT is the proportionately higher dose of radiation to which the subject is exposed. This relatively expensive equipment is available in all major hospitals.

5.1.5 Ultrasonography

Ultrasonography, an ultrasound-based diagnostic imaging technique is used to visualize muscles, tendons, and many internal organs to measure their size, structure, and any pathological lesions with real-time tomography images [50]. It offers high spatial resolution (<1 mm) and excellent anatomical and molecular information. Besides, it is relatively cheap and portable when compared with other imaging modalities [1, 50]. It has been used for around 50 years; its nonradioactive nature poses no known risks to patients, and it is widely used to visualize fetuses and emergency prenatal care [5].

A range of ultrasonography contrast agents have been designed, including microbubbles, liposomes, or perfluorocarbon emulsion. The characteristics of these contrast agents are summarized in Table 5.3. The relatively large size of the imaging agent particles (<250 nm) is a major limitation of ultrasound as molecular imaging modality, which can restrict tissue penetration, and hence limits its application for vascular targets [4].

Table 5.3 Ultrasound contrast agents

Properties, Indications	Substances	Form
Gas bubbles, short half life, difficult reproducibility, cavitation and shaking effect	Shaken saline, Indocyanine green	Solutions
High density, low acoustic velocity, lower solubility in blood, prolongs Doppler enhancement	Perfluorooctylbromide, lipid emulsions	Colloidal emulsions (form)
Aqueous microparticle suspension with very fine gas bubbles.	Iodipamide, ethyl ester, microparticles	Colloidal suspensions
Heart disease diagnostics	Galactose microparticles suspension forming microbubbles	Suspensions with gas microbubbles

Note: Adapted from [5].

5.2 Rare Earth Based Multimodal Molecular Imaging Contrast Agents

Synergistic combination of two or more imaging modalities is made possible by advanced hardware technology and the development of novel multimodal imaging probes. Integrated imaging modalities can provide more comprehensive and reliable biological data, which is important for accurate diagnosis. Taking PET/MRI and MRI/optical imaging as examples, the combination of high-sensitivity PET and high-resolution MRI can be used in brain imaging, where tissue remains relatively fixed in the skull [51]. Combination of optical imaging and MRI allows the scalpel to be guided (via fluorescence imaging) to remove lesions (via MRI) [52].

Rare earth based nanomaterials have been widely investigated as contrast agents in integrated imaging modalities in recent years. Such integrated imaging modalities include MRI/ fluorescence imaging, PET/MRI and CT/MRI. The next section is broadly divided into two sections to discuss the application of rare earth ions as contrast agents in (1) MRI/fluorescence imaging and (2) PET/MRI and CT/MRI.

5.2.1 MRI/Optical Imaging Probes and Their Bioapplications

The most investigated inorganic multimodal imaging probes combine MRI and optical imaging capabilities. In MRI, the most commonly used T_1 contrast agents are low molecular weight gadolinium ion complexes. Gadolinium ions have seven unpaired electrons, which can efficiently alter the relaxation time of water where they distribute. Superparamagnetic iron oxide nanoparticles are commercially available as T_2 contrast agents. Dysprosium (Dy) is another lanthanide ion that has been investigated in MRI, and in most cases, behaves as a T_2 contrast agent. Based on the type MRI contrast agent, rare earth based multimodal imaging probes can be generally categorized into two groups: (1) Gd^{3+}-based multifunctional nanomaterials, (2) superparamagnetic iron oxide nanoparticles based multifunctional nanomaterial and (3) Dy^{3+}-based multifunctional nanomaterials.

5.2.1.1 Gd^{3+}-based multifunctional nanomaterials

Gd ions, as paramagnetic centers, have to be incorporated into chelate ligands or doped into a matrix as MRI contrast agents due to their toxicity. The most widely used T_1 contrast agent is gadolinium diethylenetriaminepentaacetate (Gd-DTPA) [34]. To develop new types of multifunctional nanomaterials, Gd ions are incorporated into various nanostructured materials, including quantum dots (QDs), mesoporous silica materials, silica nanoparticles, and other types of fluorescent nanoparticles. The following sections highlight recent development in such materials as magnetic-fluorescent nanoparticulate probes.

5.2.1.1.1 *Gd^{3+}-quantum dots magnetic-fluorescent nanoparticles*

QDs have been proposed as type of novel and efficient fluorophores due to their numerous advantages: size-dependent emission, high brightness, narrow emission, broad absorption spectra, and high resistance to photobleaching [53]. Many studies have reported magnetic-fluorescent QDs by conjugating or doping Gd^{3+} ions directly into the QDs.

A recent study reported Gd-DOTA (1,4,7,10-tetraazacyclododecane-1,4,7,10-tetraacetic acid) complex functionalized on the surface of glutathione (GSH, in short)-coated QDs, as shown in Fig. 5.1A [54]. The labeled QDs were reported to have r_1 relaxivity of 365 $mM^{-1}s^{-1}$. However, it is based on per mole QDs, not Gd ions. Its utility as a dual modal contrast agent for fluorescence imaging (Fig. 5.1B1) and *in vivo* MRI of a mouse (Fig. 5.1B2) demonstrated that the combination of near-infrared (NIR) fluorescence and magnetic resonance images would enable facile identification of the location of target cells and/or tissues labeled with fluorophores.

Gerion *et al.* presented QDs coated with a 1–2 nm thick PEGylated silica shell. Amine reactive Gd-DOTA complexes were then cross-linked onto the silica shell using sulfosuccinimidyl-4-(*N*-maleimidomethyl)cyclohexane-1-carboxylate(sulfo-SMCC) [55]. Magnetic resonance relaxivities of the resultant Gd-DOTA-modified QDs were r_1 = 1019 $mM^{-1}s^{-1}$ (0.5 T) and r_2 = 2348 $mM^{-1}s^{-1}$ (1.4 T). Preliminary dynamic contrast enhancement MRI experiments in mice revealed that the probes were cleared from the renal system into the bladder with no observable adverse effects on the health of the animal. Prinzen *et al.* reported another QDs-based magnetic-fluorescent nanoparticles for visualization of cell death and activated platelets with fluorescence imaging and MRI [56]. A biotinylated construct consisting of a lysine wedge with Gd-DTPA complexes was conjugated to biotinylated anxA5, coupled to streptavidin-coated QDs-based nanoparticles. The molecular structures of the biotinylated Gd-DTPA (a) and biotinylated Gd-wedge (b) structures with biotin (red) and Gd-DTPA (blue) are presented by Fig. 5.2. The particles exhibit intense fluorescence and a large magnetic resonance relaxivity (r_1) of 3000–4500 $mM^{-1}s^{-1}$ per nanoparticle due to the newly designed construct which increased the Gd-DTPA loading. The nanoparticles were suitable for both anatomical and subcellular imaging of structures in the vessel wall.

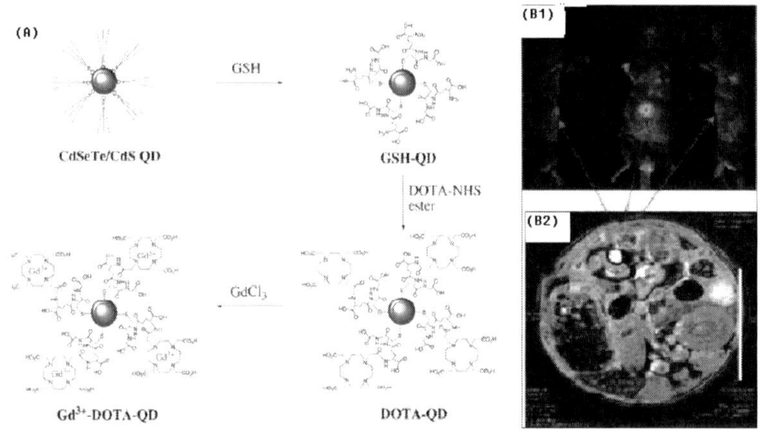

Figure 5.1 (a) Schematic representation for the preparation of Gd^{3+}-DOTA functionalized CdSeTe/CdS QDs with glutathione (GSH) coating; (b_1) a NIR-fluorescence image and (b_2) a T_1-weighted magnetic resonance axial image for a mouse. A phantom containing 10 mM of Gd^{3+}-DOTA-QDs is embedded into the mouse abdomen. The arrow heads on the red broken lines represent the location of the phantom being visualized by fluorescence imaging and MRI. The white bar in the magnetic resonance image corresponds to 1 cm in length [54].

Yang *et al.* grafted Gd (III) ions on ZnS-passivated CdS: Mn QDs [57]. *N*-(Trimethoxysilylpropyl)ethyldiamine, triacetic acid trisodium salt (TSPETE) was attached to the QDs. A yield of 107 Gd^{3+} ions per QD was reported. However, TSPETE has only five coordination sites for gadolinium compared with the eight coordination sites for DOTA and nine for DTPA. Thus it would be expected to be less stable *in vivo*. The magnetic resonance relaxivities of the QDs were r_1 = 20.5 $mM^{-1}s^{-1}$

and r_2 = 151 $mM^{-1}s^{-1}$ at 4.7 T. Mulder *et al.* reported the synthesis of QDs with a water-soluble and paramagnetic Gd-based micellular coating as molecular imaging probes for fluorescence microscopy and MRI [58]. The core/shell CdSe/ZnS QDs exhibited excellent optical properties. Coating the micellular and paramagnetic shell rendered the nanoparticles magnetic resonance-detectable, water-soluble, and biocompatible. Magnetic resonance results showed that the nanoparticles had high r_1 relaxivities, which was three times higher than that of Gd-DTPA (Fig. 5.3). Paramagnetic QDs were later conjugated with cyclic arginine–glycine–aspartic acid (RGD) peptides and were successfully targeted to human endothelial cells *in vitro*. This suggests that nanoparticulate bimodal agents may be useful for the detection of tumor angiogenesis.

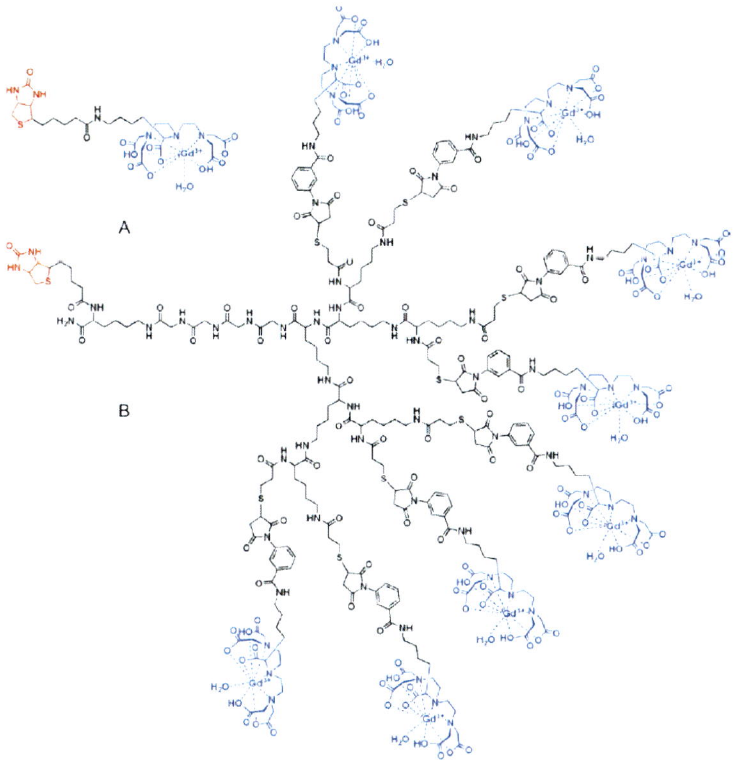

Figure 5.2 Molecular structure of the biotinylated Gd-DTPA (a) and biotinylated Gd-wedge (b) structures with biotin (red) and Gd-DTPA (blue) [56].

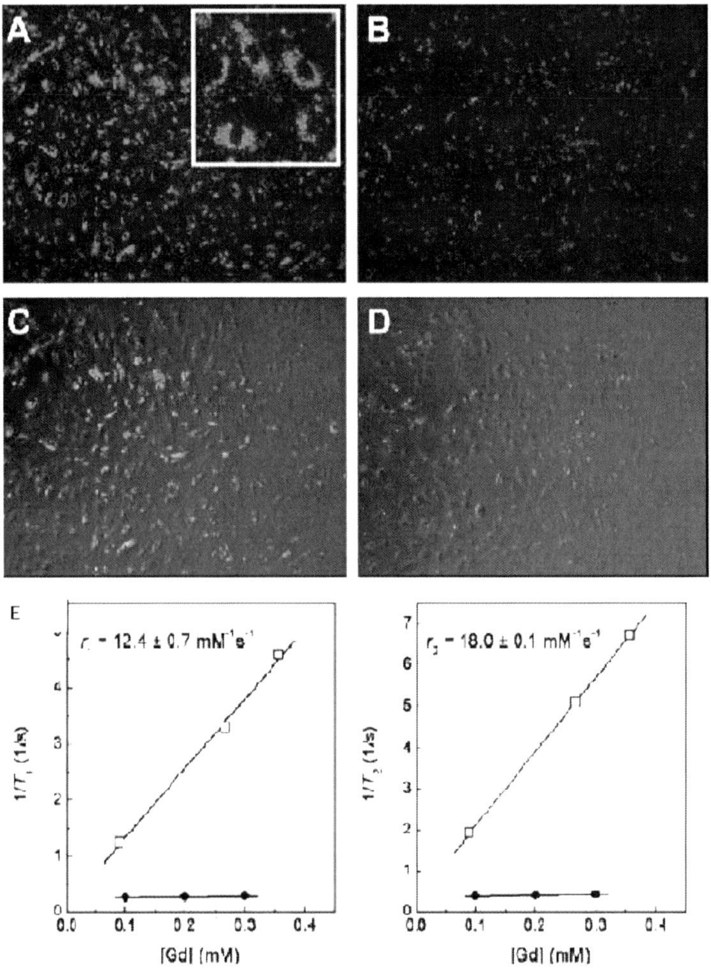

Figure 5.3 Fluorescence microscopy of HUVEC incubated with (a) RGD-pQDs and (b) bare pQDs. The corresponding phase contrast images of (a) and (b) are depicted in (c) and (d), respectively. (e) longitudinal (left) and transverse (right) relaxivity of lipid-coated QDs with (squares) and without (circles) paramagnetic lipids in the micellular coating [58].

5.2.1.1.2 *Magnetic-fluorescent silica and mesoporous silica nanoparticles*

The accessibility of the magnetic centers to water molecules is vital for highly efficient T_1 contrast agents. Mesoporous materials provide an ideal platform for the development of magnetic resonance–enhancing hybrid materials due to their high surface areas, tunable pore structure, rigid structure, larger pore volume, great surface-modification capability, and good biocompatibility.

Tsai *et al.* fabricated paragmagnetic Gd^{3+}-dye@mesoporous silica nanorods (MSN-R) for multifunctional cell imaging [59]. Dye@ MSN-R nanoparticles were first synthesized in the presence of the surfactant cetyltrimethylammonium bromide (CTAB). After the removal of the surfactant, the nanorods reacted with $GdCl_3{\cdot}6H_2O$ to form the multifunctional nanorods Gd-Dye@ MSN-R. The nanorods displayed relaxivities with r_1 = 22 $mM^{-1}s^{-1}$ and r_2 = 41 $mM^{-1}s^{-1}$ at 0.47 T, which were about five and ten times higher than the values of the complex $[Gd(DTPA)]^{2-}$. The low r_2: r_1 ratio suggested that Gd-dye@ MSN-R served as a good dual-contrast agent for both T_1- and T_2-weighted MRI. Cell viability and cell function were performed to examine its function as a cell marker, and there was no adverse effect on cell viability. High nanorod uptake by the cells was observed without the use of any transfection agent.

Hsiao *et al.* reported another type of mesoporous silica nanoparticles as a delivery system of gadolinium for effective human stem cell tracking [60]. Gd-fluorescein isothiocyanate mesoporous silica nanoparticles (Gd-Dye@MSN) possessed both green fluorescence and paramagnetism. Human mesenchymal stem cells (hMSCs) were labeled with Gd-Dye@MSN via endocytosis. An MRI was undertaken by a clinical 1.5 T MRI system, and a low incubation dosage of Gd, low detection cell numbers, and short incubation times were demonstrated on both loaded cells and hMSC-injected mouse brain. The result indicated that MSN may be an ideal vector of T_1 agent for stem cell tracking with MRI. In another example, a different Gd-Si-DTTA complex was grafted onto MSNs via siloxane linkage [61]. The magnetic resonance relaxivities of the particles were measured using both 3.0 T and 9.4 T magnetic resonance scanners. On a per millimolar Gd basis, the particles showed r_1 values of 28.8 $mM^{-1}s^{-1}$ at 3 T and 10.2 $mM^{-1}s^{-1}$ at 9.4 T, while r_2 relaxivities were 65.5 $mM^{-1}s^{-1}$ at 3 T and 110.8 $mM^{-1}s^{-1}$ at 9.4 T, respectively. They attributed the relative larger relaxivities to the ready access of water molecules through the nanochannels of the MSN-Gd particles. Their utility as contrast agent for optical imaging and MRI were also demonstrated *in vitro*. MSN-Gd showed a highly efficient T_1 contrast effect for intravascular MRI and an excellent T_2 contrast effect of soft tissue when applied at a higher dosage (Fig. 5.4).

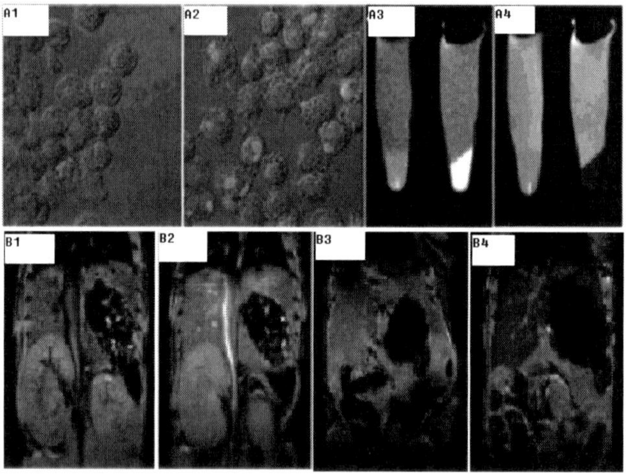

Figure 5.4 Upper row: Overlaid DIC and fluorescence image of monocyte cells incubated with (a_1) no MSN-Gd-1 and (a_2) 4 g of MSN-Gd-1 (cells in 2 mL medium) (60 nm magnification). (a_3) T_1-weighted magnetic resonance images of monocyte cell pellets incubated without MSN-Gd (left) and with 0.3 mg of MSN-Gd for cells in 3 mL of media (right). (a_4) T_2-weighted magnetic resonance images of monocyte cell pellets incubated without MSN-Gd (left) and with 0.3 mg of MSN-Gd for cells in 3 mL of media (right). Bottom row: (b_1) Pre-contrast and (b_2) post-contrast T_1-weighted mouse magnetic resonance image showing aorta signal enhancement. (b_3) Pre-contrast and (b_4) post-contrast mouse magnetic resonance images showing liver signal loss due to T_2-weighted enhancement [61].

Rieter *et al.* [48] synthesized hybrid silica nanoparticles, which contained a luminescent $[Ru(bpy)_3]Cl_2$ core (bpy = 2,2'-bypyridine) and a paramagnetic silylated Gd complex monolayer by using a reverse microemulsion method, as shown in Fig. 5.5a. In order to enhance the Gd^{3+} loading capacity, mono- and bis-silylated Gd complexes were synthesized and loaded onto the Ru complex-silica cores, respectively. Magnetic resonance relaxivities showed that this type of nanoparticles had an r_1 of 19.7 $mM^{-1}s^{-1}$ and an r_2 of 60.0 $mM^{-1}s^{-1}$ on a per milimolar Gd^{3+}-ion. Monocyte cells were chosen for the *in vitro* study because of their phagocytic capacity as well as their important role in autoimmune diseases. Laser scanning confocal study showed that the cells had been efficiently labeled with the hybrid nanoparticles to allow multimodal *in vitro* imaging. Positive and negative signal enhancement in the T_1-weighted images were both observed for the labeled cells, depending on the magnetic resonance pulse sequence employed, as shown in Fig 5.5(2c, 2d). Cell viability showed that the nanoparticles were not toxic to cells (Fig 5.5(2f)).

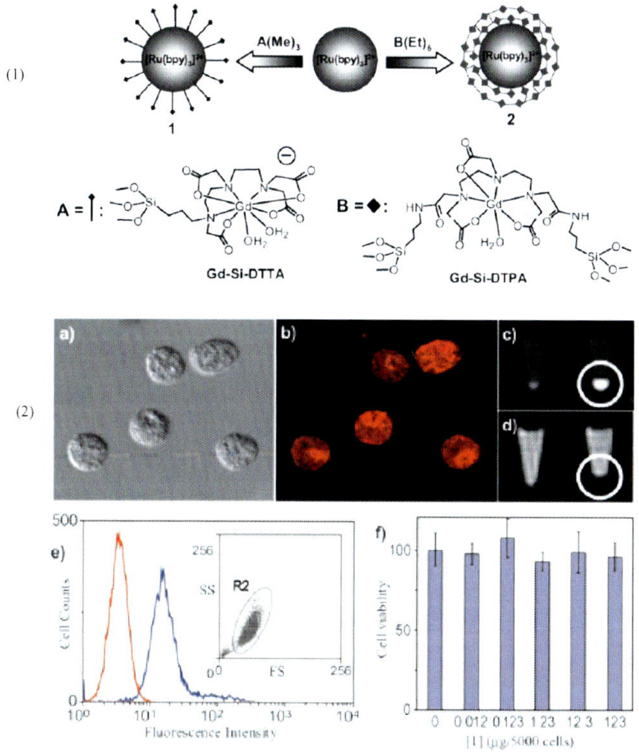

Figure 5.5 (1) Schematic representation for the preparation of the hybrid dye-doped silica nanoparticles (2) Microscopic images of 1-labeled monocyte cells: (a) optical; (b) laser scanning confocal fluorescence. (c–d) MR images of unlabeled (left) and 1-labeled (right) monocyte cells: (c) T_1-weighted and d) T_2-weighted. (e) Flow cytometric results for the unlabeled (red) and 1-labeled (blue) monocyte cells indicating greater than 98% labeling efficiency (the inset shows the purity of the labeled cells; SS = side scatter, FS = forward scatter). (f) MTS assay of the monocyte cells incubated with different amounts of nanoparticles [48].

For MRI/optical probes, incorporating high payload magnetic centers for detection is more desirable due to the relatively low sensitivity of current clinical MRI. Therefore, the same group increased the relaxivities of Gd^{3+}-chelated dye-doped silica nanoparticles using layer-by-layer polyelectrolyte deposition method [62]. The Gd-DOTA (DOTA = 1,4,7,10-Teraazactckididecabe-1,4,7,10-Tetraacetic acid) complex was deposited on the Gd-Si-DTTA (Gd-(siloxylpropyl)diethylenetriamine tetraacetate (Gd-DTTA))-modified silica nanoparticle via the electrostatic interaction between the positively charged complex and the negatively charged polystyrenesulfonate (PSS) group (Fig. 5.6(1)). The r_1 and r_2 values per particle were increased proportionally with the increased number of layers (Fig 5.6(2a, 2b)). This indicated that water molecules could readily access the Gd centers, thus allowing efficient water proton relaxivity. The *in vitro* imaging of cancer-specific magnetic-fluorescent nanoparticles was also demonstrated by the noncovalent functionalized positive peptides via electrostatic interaction (Fig 5.6(2c–2i)).

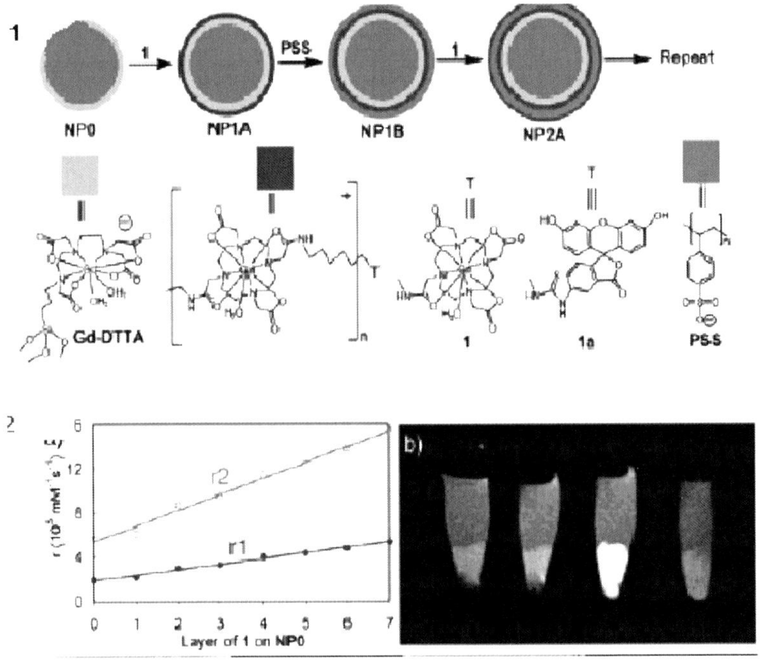

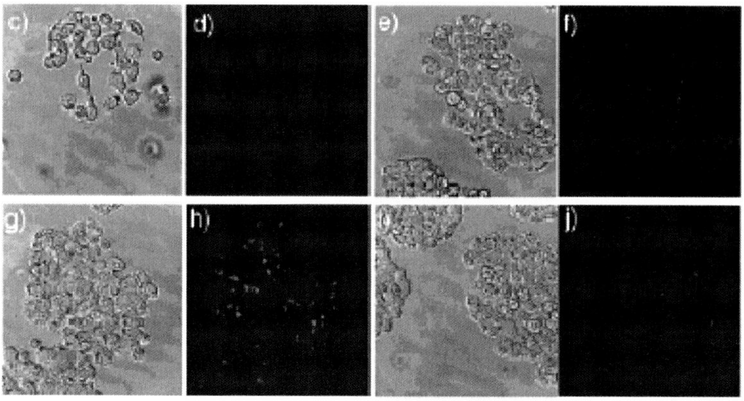

Figure 5.6 (1) Schematic representation for the preparation of the hybrid dye-doped silica nanoparticle. (2) (a) dependence of per particle r_1 and r_2 values on the number of deposited Gd-DOTA oligomer 1 on NPO. (b) T_1-weighted magnetic resonance images of HT-29 cells that have been incubated with various nanoparticles. From left to right: control cells without any nanoparticle, cells with NP3B particles. Cells with NP3B particles that have been noncovalently functionalized with K_7RGD and cells with NP3B particles that have been noncovalently functionalized with K_7GRD, (c–j) phase contrast optical (c, e, g, and i) and LSCFM microscopic images (d, f, h, and j) of HT-29 cells that have been incubated with various nanoparticle: control cells without any particles (c and d), cells with NP5B particles (e and f), cells with NP5B particles that have been noncovalently functionalized with K_7RGD (g and h), and cells with NP5B particles that have been noncovalently functionalized with K_7GRD (i) [62].

Santra *et al.* also reported multifunctional single-core multiple-shell Ru(bpy):Gd(III)/SiO_2 nanoparticles, which were photostable, radio-opaque, and paramagnetic [63]. Two silica layers were coated onto the silica core, sequentially namely, hydrated inner silica shell and outer silica shell. Highly sensitive and photostable fluorescent dye Ru(bpy) was first encapsulated into silica core and paramagnetic Gd ions were then bound to TSPETE in the hydrated inner silica shell via the microemulsion method. Another silica shell was then coated on with primary amino groups for bioconjugation. Dye molecules gave orange-red fluorescence by UV excitation, and magnetic resonance data showed efficient water exchange with the Gd^{3+} ions, although they were embedded inside the amorphous silica shell.

5.2.1.1.3 *Gd-doped and Gd-based fluorescent nanomaterial*

Surface grafting processes usually result in a low payload of Gd^{3+} complex on a single nanoparticle. This limits its application in localized targeting imaging. Co-doping Gd^{3+} inside fluorescent nanocrystals is a more straightforward and effective route to obtain a high payload of Gd^{3+} multifunctional nanocrystals for dual modal imaging. Li *et al.* synthesized dual modal optical and MRI imaging agent CdSe: Gd nanoparticles [64] using mercaptopropylic acid (MPA) as a capping agent. MPA also rendered the nanoparticles water soluble. To enhance the fluorescence quantum yield as well as water dispersity, the MPA-capped QDs were further coated with lysine molecules (Fig. 5.7(1)). Confocal laser images showed that orange luminescent lysine-capped CdSe: Gd NPs were included inside the vero cells (Fig. 5.7(2a, 2b). T_1 relaxation showed that the CdSe: Gd NPs had much larger r_1 value than the conventional Gd-DTPA (Fig. 5.7(3).

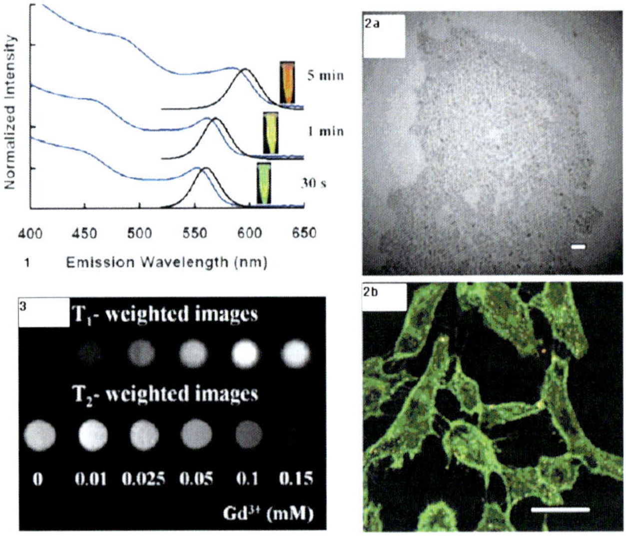

Figure 5.7 (1) Absorption and photoluminescence spectra of CdSe: Gd nanoparticles at different reaction times (absorption: blue; emission: black; emission spectra excited at 450 nm). (2a) TEM images of lysine-capped CdSe: Gd nanoparticles (scale bar = 20 nm); (2b) confocal laser imaging of vero cells treated by nanoparticles (scale bar = 50 μm). (3) MRI *in vitro* assays of T_1- and T_2-weighted images of lysine-capped CdSe: Gd NPs [64].

Some lanthanide oxides have been shown to be promising as multimodal contrast agents. Examples include Gd_2O_3 encapsulated with a layer of organic fluorophores [65], gadolinium oxides/carbon hollow shells [69], Gd_2O_3 nanoparticles embedded within single-walled carbon nanohorn (SWNHs) [70], Gd_2O_3 ultrasmall nanorods doped with lanthanide ions [66, 67] and Y_2O_3 doped with Gd^{3+} ions [68].

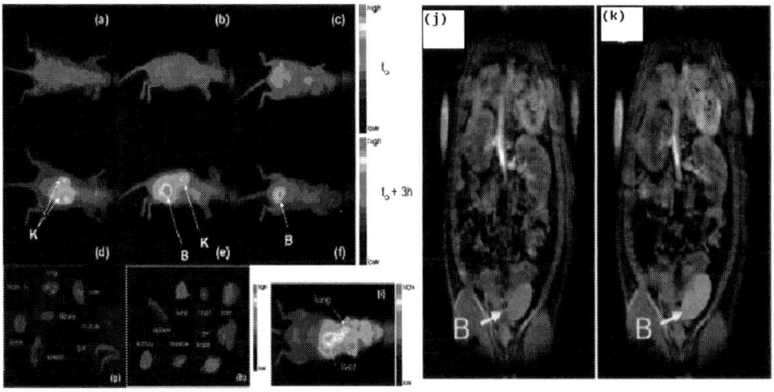

Figure 5.8 Fluorescence reflectance imaging of a nude mouse (a, b, c) before and (d, e, f) 3 h after the injection of GadoSiPEG2C (K, kidneys; B, bladder). Fluorescence reflectance imaging of some organs after dissection (g) of a control mouse (no particles injection) and (h) of the nude mouse visualized on pictures (a–f). (i) Fluorescence reflectance imaging of a nude mouse after the injection of GadoSi2C (particles without PEG). Each image is acquired with an exposure time of 200 ms. T_1-weighted images of rat (j) before and (k) 1 h after injection of GadoSiPEG2C (B for bladder) [65].

Gd_2O_3 nanoparticles may find potential applications in optical detection, bimodal imaging, and therapeutics. Bridot *et al.* pioneered the work of Gd_2O_3 nanoparticles as multimodal contrast agents [65]. In their work, ultrasmall Gd_2O_3 core was embedded in a polysiloxane shell whose outer part was functionalized with PEG and inner part by organic dyes. Gadolinium oxide core induced an enhancement in T_1-weighted MRI, and the organic dyes produced fluorescence. Relaxivities characterization revealed that relaxivities decreased with increasing core size. Thus, the 2.2 nm Gd_2O_3 core showed greatest contrast in MRI. This is due to decreasing surface/volume ratio as the smaller particles present the greatest number of gadolinium ions at the surface of the same concentration. After intravenous injection of the nanoparticles in female Swiss nude mice, *in vivo* fluorescence imaging showed that the nanoparticles (Fig. 5.8a–i) were accumulated in the kidney and bladder. The *in vivo* T_1-weighted MRI images substantiated the fluorescence imaging results, where nanoparticles were found to accumulate in bladder (Fig. 5.8j,k).

Bifunctional Gd_2O_3/C hollow nanoshells for application in magnetic resonance contrast imaging and NIR fluorescence have been developed by Huang *et al.* [69]. The nanoshells were prepared using biological gelatin as core templates through a two-step thermal treatment, and 384 nm gelatin templates yielded a carbon shell of 183 nm in diameter with a thickness of 19.2 nm. Poly(styrene-alt-maleic acid) (PSMA) was used as a surfactant to modify the surface of the nanoshell as well as improve their water dispersity and increase their biocompatibility (Fig. 5.9). Interestingly, the Gd_2O_3/C nanoshells showed brightened images of kidney cortex and liver in mice, whereas the Gd_2O_3/C@PSMA nanoshells showed darken liver signal (Fig. 5.10). Cytotoxicity assay showed that both Gd_2O_3/C and Gd_2O_3/C@PSMA nanoshells had no significant toxicity and could be eliminated after 24 h. The graphite carbon coated Gd_2O_3 nanoshells displayed a large extinction coefficiency at 808 nm with good absorbance in the NIR region. Therefore, the Gd_2O_3/C nanoshell can be used as photothermal therapeutic agents. By conjugating anti-EGFR antibodies (humanized anti-epithermal growth factor receptor (EGFR)), the nanoshells can effectively target malignant A549 lung cancer cells and kill them by photothermolysis (Fig. 5.11). These nanoparticles offered possibilities for loading drugs or other imaging functionality. However, particles of smaller sizes will be required for *in vivo* work.

Figure 5.9 Schematic representation for the synthesis of Gd_2O_3/C@PSMA nanoshells and their biomedical applications [69].

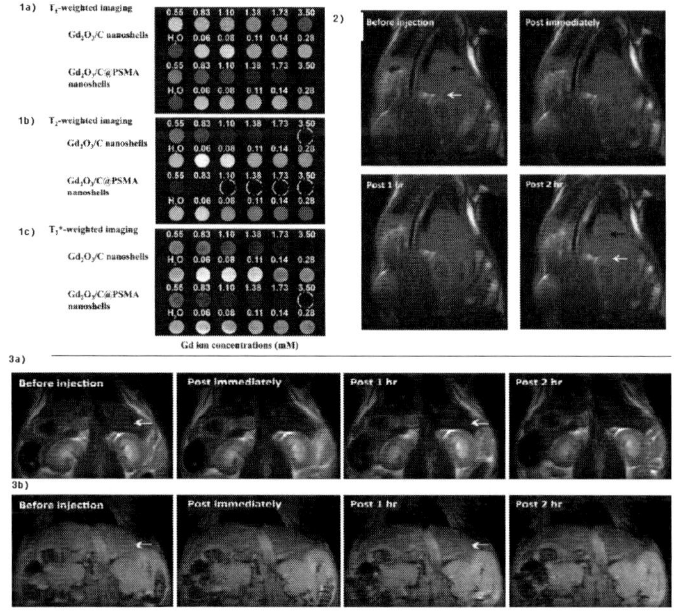

Figure 5.10 MRI images of Gd_2O_3/C nanoshells and Gd_2O_3/C@PSMA nanoshells. (1a) T_1-weighted imaging; (1b) T_2-weighted imaging; and (1c) T_2*-weighted imaging. (2) *In vivo* progressive MRI events. T_1-weighted images of male BALB/c mice administrated with Gd_2O_3/C nanoshells at the indicated temporal points (pre-injection, immediate post, 1 h, and 2 h) (The white and black arrows indicate the kidneys and liver, respectively). (3) *In vivo* progressive magnetic resonance images of Gd_2O_3/C@PSMA nanoshells: (3a) T_2-weighted and (3b) T_2-weighted images of male BALB/c mice at the temporal points (pre-injection, immediate post, 1 h, and 2 h). The white arrows indicate the liver [69].

In another example, Gd_2O_3 nanoparticles were embedded within single-walled carbon nanohorn (SWNHs), which were used to quantitatively determine biodistribution and structural localization of drug delivery systems [70]. Gd_2O_3@SWNH aggregates were intravenously injected into mice, and 70–80% of the total injected Gd_2O_3@SWNH aggregates accumulated in the liver, and about 10% in the spleen. Das *et al.* reported a simple strategy for synthesizing ultranarrow gadolinium oxide nanorods as multimodal contrast agents [71]. Downconversion and upconversion fluorescence can

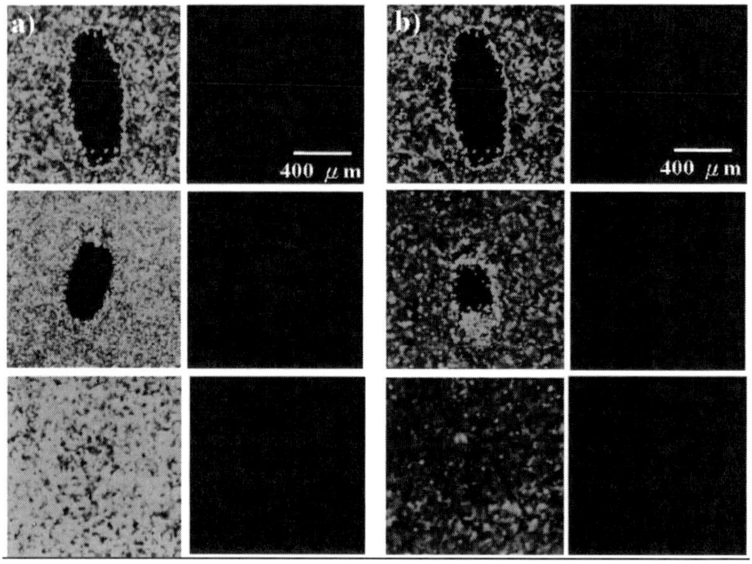

Figure 5.11 (a) Anti-EGFR conjugated with Gd_2O_3/C@PSMA nanoshells (500 mgmL^{-1}) treated with A549 cancer cells were irradiated by laser dosages of 20 Wcm^{-2} (top row), 15 Wcm^{-2} (middle row), and 10 Wcm^{-2} (bottom row) for 7 min. (b) Anti-EGFR conjugated with Gd_2O_3/C@PSMA nanoshells irradiated by the laser dosages of 20 Wcm^{-2} as a function of nanoshell dosages: 500 mgmL^{-1} (top row), 400 mgmL^{-1} (middle row), and 300 mgmL^{-1} (bottom row). The column shows staining by calcein AM as green fluorescence dye for living cells. The column displays staining by EthD-1 as red fluorescence for dead cells [69].

be achieved by changing the lanthanide dopants, as shown in Fig. 5.12a,b. Besides, the Yb/Er co-doped Gd_2O_3 nanoparticles exhibited good T_1-weighted MRI contrast, comparable to the commercial product Gadovist^{TM}.

In addition to bimodal imaging applications, Setua *et al.* demonstrated targeting of cancer cells (folate receptor positive human nasopharyngeal carcinoma cells (KB)) by Eu^{3+} and Gd^{3+} ions doped Y_2O_3 nanocrystals [68]. The rare earth doped nanoparticles were then conjugated with folic acid to target the KB cells. The study presented *in vitro* cell targeting and imaging by those bimodal nanoparticles (Fig. 5.13a,b) while *in vitro* MRI images of the nanoparticles showed good MRI contrast (Fig. 5.13c).

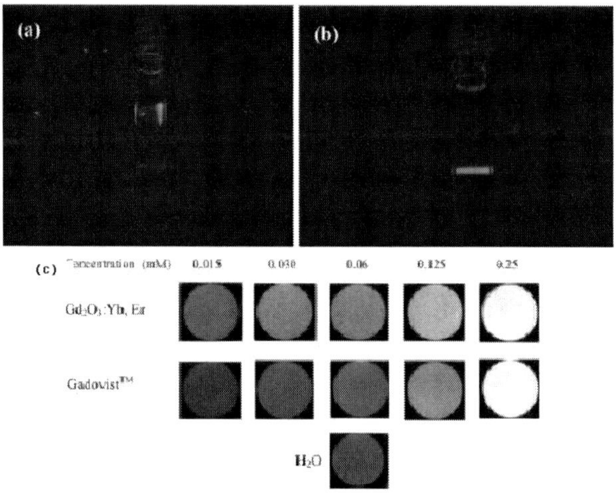

Figure 5.12 Digital photographs of (a) downconversion Tb-doped Gd_2O_3 solution exhibiting green emission at 235 nm excitation and (b) upconversion Yb/Er-co-doped Gd_2O_3 solution exhibiting red emission at 980 nm excitation. (c) T_1-weighted images of Gd_2O_3:Yb, Er and Gadovist™ at various Gd^{3+}concentrations. T_1-weighted image of water sample is shown as reference [71].

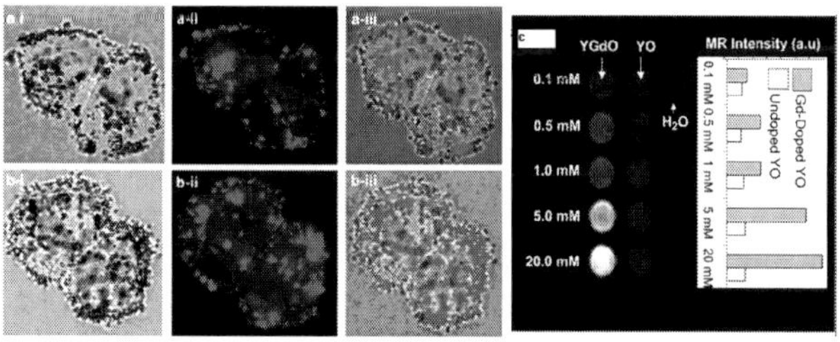

Figure 5.13 Fluorescent microscopic images (a–i, b–i-phase contrast, a–ii, b–ii-dark-field fluorescence and a–iii, b–iii-bright-field fluorescence) showing interaction of FA-PEI-YO NCs with FRt KB cells: Imaging experiment carried out with 2 mM red emitting $Y_{1.97}Gd_{0.03}O_3$:2%Eu^{3+} NCs. Imaging after 2 h of incubation showing specific accumulation of large concentration of YO NCs on the cell membrane of FRt KB cells. MRI contrast imaging of doped (2% Eu, 0.03% Gd) and undoped nanoparticles of different concentrations using a clinical imaging system (GE Healthcare) (c). The imaging parameters include T_1-weighted spin echo pulse sequences at 1.5 T with the following parameters: TR/TE = 200 ms/9 ms, slice thickness of 0.6 mm, 256*256 points and number of acquisitions = 4. (Inset) Relative MRI signal intensity of undoped and doped NCs with increasing Gd concentration [68].

Fluoride host materials offer distinct advantages over oxide materials, as the fluoride matrix has low absorbency and the emitter can be excited directly. The quenching of excited state rare earth ions is minimized when lanthanide ions are doped into fluoride hosts, which leads to long emission lifetimes and high luminescence intensities even in the case of IR-emitting ion [72]. Wong *et al.* investigated the magnetic and luminescent properties of multifunctional GdF_3: Eu nanoparticles [73]. The nanoparticles showed downconversion and exhibited paramagnetism at both 293 and 77 k. They later doped upconversion emitting ions Yb, Tm in the same host nanoparticles [74]. The GdF_3: Yb, Tm nanoparticles exhibited nearly pure NIR-to-NIR upconversion, which achieved deeper tissue penetration and reduced autofluorescence from biological tissues. The nanoparticles also showed paramagnetism due to the presence of Gd ion. This indicated that these nanoparticles were promising in bioimaging and bioseparation. Kumar *et al.* reported lanthanide ion doped $NaYF_4$ nanocrystals for optical and magnetic resonance bioimaging [75]. Lanthanide ions Gd^{3+} and $Er^{3+}/Yb^{3+}/Eu^{3+}$ were co-doped into fluoride nanoparticles, and the nanoparticles were then modified with tumor-specific antibodies. The *in vitro* targeted imaging of live cancer cells showed no sign of cellular toxicity. Gd^{3+} co-doped nanoparticles imparted strong T_1 contrast effect and strong upconversion fluorescence, as shown in Fig. 5.14. Thus these fluoride nanoparticles doped with lanthanide ions may be used as dual-modal contrast agents in optical imaging and MRI.

Yet in another study, by replacing yttrium (Y) with paramagnetic gadolium (Gd), Heyon *et al.* reported uniform and monodisperse Yb/Er co-doped $NaGdF_4$: Yb^{3+}, Er^{3+} nanocrystal as bimodal contrast agent [77]. The core $NaGdF_4$: Yb, Er upconversion nanoparticles were first synthesized by a thermal decomposition method, and an inert $NaGdF_4$ layer was further coated protection to provide for the doped absorption/emission centers (Yb^{3+} and Er^{3+}). The nanoparticles were found to be extremely resistant to photobleaching, and no photoblinking occurred. Good T_1-weighted MRI contrast was obtained in 1% agarose solution and in cells (SK-BR-3) incubated with nanoparticles (Fig. 5.15a,b). Besides, *in vitro* upconversion cellular imaging data (in SK-BR-3 cells) revealed cellular uptake of nanoparticles with a complete absence of autofluorescence at NIR (980 nm) excitation (Fig. 5.15c–f).

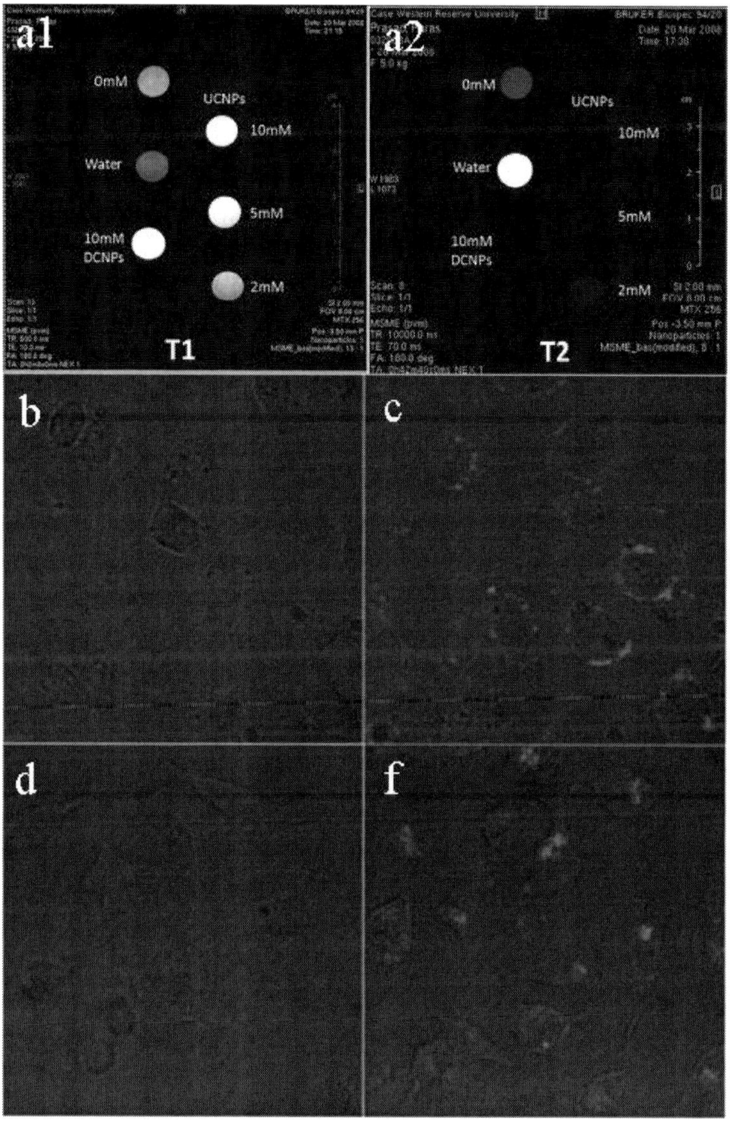

Figure 5.14 T_1- and T_2-weighted images of $NaYF_4$ doped with different concentrations of Gd and Er/Yb (UCNPs) and Eu (DCNPs) (a1–2); Confocal images of Panc 1 cells treated with UCNPs ($NaYF_4$: 2% Er, 10% Yb, 10% Gd; c, d) and DCNPs ($NaYF_4$: 10% Eu, 10% Gd.; (e, f). (b) and (d) show nonbioconjugated nanophosphors; (c) and (f) show cells targeted with nanophosphors conjugated with anticlaudin and antimesothelin, respectively [76].

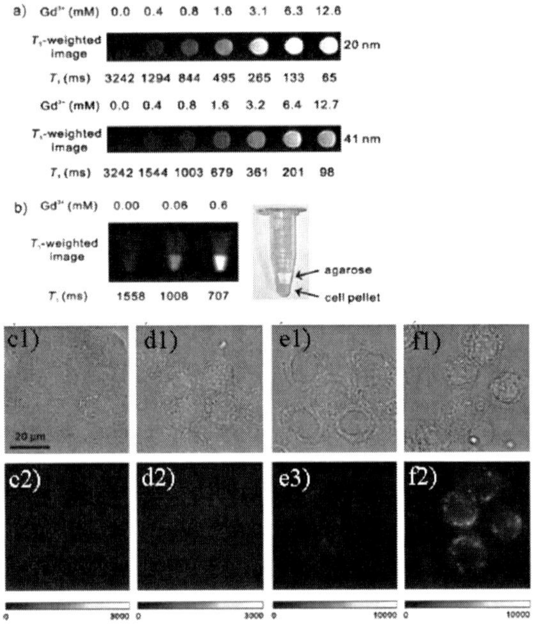

Figure 5.15 (a) T_1-weighted magnetic resonance images of UCNPs with edge lengths of 20 nm and 41 nm at various concentrations in water. (b) T_1-weighted magnetic resonance images of SK-BR-3 cells ($1.3*10^7$ cells) incubated with 20 nm UCNPs at various concentrations for 24 h; cellular uptake of 20 nm UCNPs in SK-BR-3 cells. (c1, d1, e1, f1) Bright field images. (c2, d2, e2, f2) Luminescence images with the excitation at 980 nm and the detection at 400–700 nm. Cells were incubated with the UCNPs with 100 mg mL^{-1} of Gd^{3+} concentration at 37.82. (c) Cells incubated without UCNPs, showing no autofluorescence. Incubation time was (d) 30 min, (e) 2 h, (f) 4 h [77].

Though lanthanide-fluoride bimodal nanocrystals are considered promising in combined optical imaging and MRI, most investigations were undertaken *in vitro*. Zhou *et al.* presented bimodal *in vivo* animals imaging using a $Tm^{3+}/Er^{3+}/Yb^{3+}$ co-doped $NaGdF_4$ nanoparticles [78]. Their studies revealed that Gd^{3+} ions at or near the surface of the nanocrystals were the primary contributor to relaxivities. Thus they selected $NaGdF_4$ as host which produced better MRI contrast than $NaYF_4$ host doped with Gd^{3+} ions. Figure 5.16 shows that upconversion luminescence signal was observed 40 min after intravenous injection. *Ex-vivo* imaging suggested that the signal mainly came from liver and spleen. *In vivo* MRI substantiated that uptake in the lung, kidney, and heart was quite insignificant and that the nanoparticles could circulate without aggregation and gradually cleared from reticuloendothelial systems such as liver and spleen where the MRI contrasts manifested. The treated mice survived more than a month without weight loss, abnormal behavior, and change in reproductive capacity, which suggested low or no toxic effects of the materials.

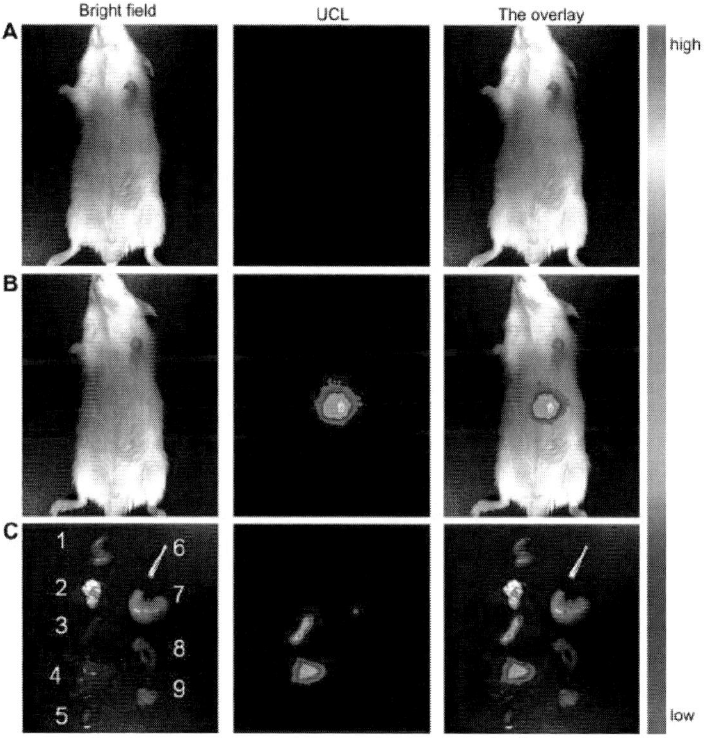

Figure 5.16 *In vivo* upconversion luminescence imaging of mouse after intravenous injection (A) with and (B) without AA-NPs. All images were acquired under the same instrumental conditions (power density = 150 $mWcm^{-2}$ on the surface of mouse). (C) Representative UCL images of dissected organs of a mouse sacrificed 40 min after intravenous injection with AA-NPs: 1, kidney; 2, lung; 3, spleen; 4, liver; 5, heart; 6, bone; 7, stomach; 8, intestines; 9, muscle. All images were acquired under the same instrumental conditions (power density = 100 $mWcm^{-2}$ on the surface of mouse) [78].

Li *et al.* reported a hybrid lanthanide nanoparticle consisting of an upconverting lanthanide-doped $NaYF_4$ nanocrystal core and a Gd-based paramagnetic complex shell [79]. The uniform core–shell structured $NaYF_4$: Yb, Er@Si-DTTA (3-aminopropyl(trimethoxysilyl) diethylenetriamine tetraacetic acid) nanoparticles were first prepared, and Gd ions were then loaded into the nanoparticles. The core/shell nanoparticles had strong fluorescence intensity and remained stable in water over seven days. Magnetic resonance relaxivities showed that the nanoparticles had r_1 of 20.1 $mM^{-1}s^{-1}$ and r_2 of 55.0 $mM^{-1}s^{-1}$ on a per milimolar Gd^{3+}-ion basis. Besides, magnetic resonance signal intensity increased along with increasing concentration of nanoparticles in the T_1-weighted mode while signal became weaker with increasing concentration of nanoparticles in the T_2-weighted mode, which suggested that the nanoparticles may be useful for both T_1- and T_2- weighted MRI.

5.2.1.2 SPIO-based multimodal imaging agents

Superparamagnetic iron oxide nanoparticles (SPIO) are representatives of T_2 MRI contrast agents. Doped with different rare earth elements or coated with a shell of rare earth elements, the "two-in-one" bifunctional nanoparticles are T_2 as well as fluorescent contrast agents. Yu *et al.* reported water-soluble Fe_3O_4@LaF_3: Ce, Tb nanocrystals with an average diameter around 30 nm [80]. The Fe_3O_4 nanoparticles were prepared by a hydrothermal method. After fabricating the core nanoparticles, they were further reacted with cetyltrimethylammnoium bromide (CTAB) to attach hydrophilic ligands to the surface of the nanoparticles, rendering them water-soluble. The CTAB-treated Fe_3O_4 nanoparticles were further coated with a fluorescent lanthanide ion doped layer (LaF_3: Ce, Tb). These nanoparticles showed strong downconversion luminescence and good magnetism. Zhang *et al.* reported single-phase Tb-doped γ-Fe_2O_3 nanocrystals, which combined superparamagnetism and luminescence into the same nanocrystals (Fig. 5.17a,b) [81]. The Tb-doped γ-Fe_2O_3 nanocrystals exhibited green photoluminescence under 235 nm excitation (Fig. 5.17c). These nanoparticles showed minimal toxicity to the cells.

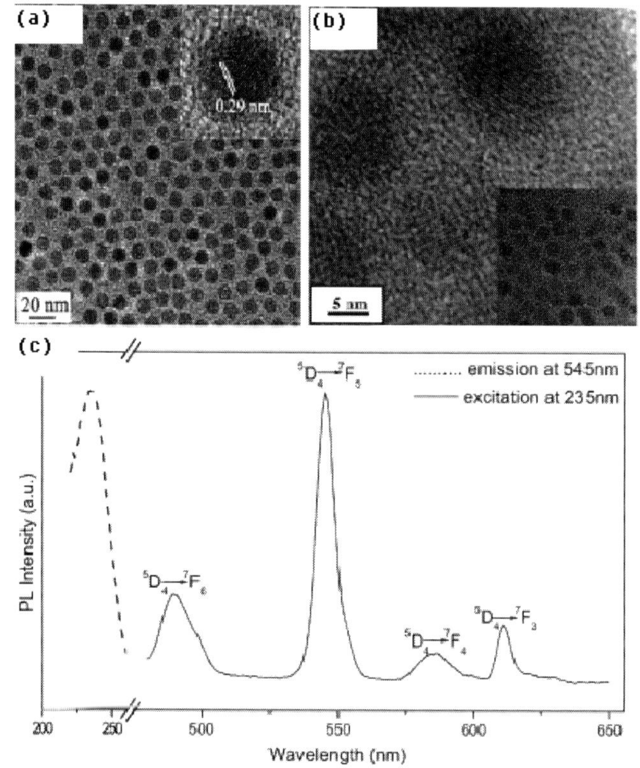

Figure 5.17 (a) A typical TEM image of Tb-doped g-Fe_2O_3 nanocrystals; inset showing HRTEM image of a single nanocrystal. (b) HRTEM image of amine-functionalized Tb-doped g-Fe_2O_3 nanocrystals; inset showing well-dispersed amine-functionalized nanocrystals in water. (c) Excitation and emission spectra of Tb-doped γ-Fe_2O_3 nanocrystals in hexane at room temperature [81].

Vaccaro *et al.* synthesized dextran-coated europium and terbium-doped ultrasmall iron oxides nanoparticles [38]. The presence of lanthanides in the iron oxide matrix conferred attractive optical properties for long-term multilabeling studies. Besides, they offered high photostability, narrow-emission band, and broad-absorption band. Furthermore, they combined high sensitivity of time-resolved fluorescence with high spatial resolution of MRI. However, most of these types bifunctional nanoparticles synthesized to date are composed of downconversion phosphors and iron oxide nanoparticles. Downconversion fluorescence has some intrinsic limitations, such as autofluorescence, low penetration depth, and phototoxicity to the biological specimen which are useless for deep tissue and *in vivo* imaging. Yet, in another study. Cheng *et al.* reported iron oxide core coated with $NaYF_4$: Yb, Er shell nanoparticles with magnetic, upconversion fluorescent, and bioaffinity properties [82]. Core/shell nanoparticles were synthesized by co-precipitation method. After coating a thin silica layer on the hybrid particles, streptavidin was immobilized on the particles, and these streptavidin-coated magnetic-fluorescent particles can bind specifically to biotinylated IgG, which showed great bioaffinity. However, no MRI test was performed.

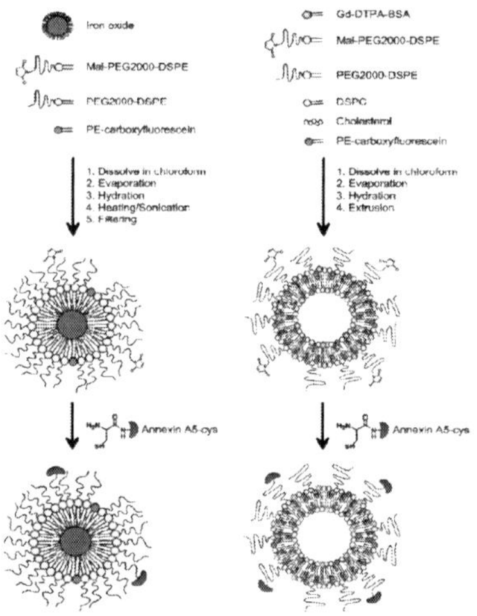

Figure 5.18 Preparation of annexin A5-functionalized superparamagnetic micelles (left) and annexin A5-functionalized paramagnetic liposomes (right). In both systems, fluorescent lipids were incorporated in the lipid layer for fluorescence microscopy or flow cytometry. Magnetic resonance imaging contrast was obtained by incorporation of superparamagnetic iron oxide in the micellular core (left) or by incorporation of paramagnetic Gd-DTPA-BSA in the lipid bilayer of the liposomes (right). Both contrast agents were functionalized with annexin A5 [83].

Van Tiborg *et al.* reported another two types of lipid-based magnetic contrast agents: gadolinium and fluorescent entities combined into a bilipid layer and hydrophobic iron oxide nanoparticles incapsulated in a fluorescent lipid-containing micellular shell (Fig. 5.18), which were aimed to detect programmed cell death with both MRI and optical techniques [83]. Amphiphiles nanoparticles with functional head-groups covalently coupled to annexin A5 proteins for specific targeting of apoptotic cells. Magnetic resonance imaging contrast was provided either by entrapment of iron oxide nanoparticles within PEGylated micelles or by incorporation of Gd-DTPA-bis(stearylamide) (Gd-DTPA-BSA) lipids within the lipid bilayer of PEGylated liposomes. Both annexin-A5 conjugated contrast agents showed significantly increasing relaxivities of apoptotic cell compared to untreated control cells and apoptotic cells that were treated with nonfunctionalized nanoparticles. Increased relaxation rates were originated from association of the contrast agents to apoptotic cells (Fig. 5.19).

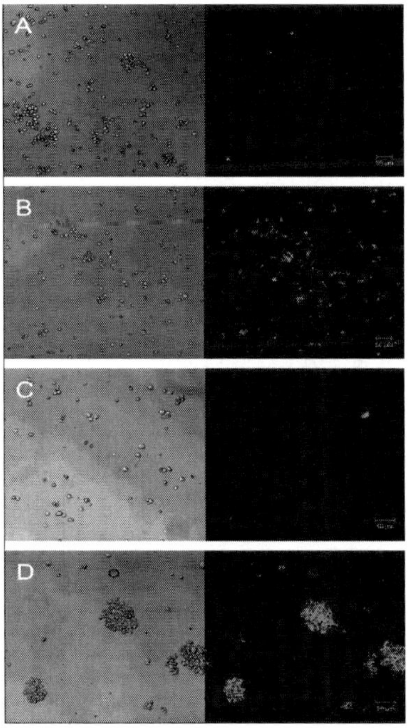

Figure 5.19 Confocal transmission (left) and fluorescence images (right) of fixed apoptotic Jurkat cells after incubation with superparamagnetic micelles (A, B) or paramagnetic liposomes (C, D) for 30 min in the presence of Ca^{2+}. Images were acquired after treatment with both nonfunctionalized (A, C) and annexin A5-functionalized (B, D) lipid-based contrast agents [83].

Streptavidin bounded upconverting $NaYF_4$: Yb, Er were deposited onto iron oxide nanoparticles in the presence of EDTA [84]. These core/shell nanoparticles emitted at 539 nm and 658 nm when excited with a 980 nm laser. The hybrid particles were found to stack in chainlike assemblies when subjected to an external magnetic field. They bound specifically to a glass slide spotted with biotinylated IgG and showed upconversion fluorescence, confirming the successful binding with the protein.

5.2.1.3 Dy-based multifunctional nanomaterials

Dysprosium is another lanthanide ion that has been investigated as a T_2 contrast agents. The present trend in clinical MRI is towards higher magnetic field strength for a better signal-to-noise ratio, enhanced resolutions as well as larger acquisition rates [85]. The current commercial Gd(III)-chelates contrast agents exhibited poor water relaxivity, especially under high field. As a result, the interests for Dy-based complex or Dy-based nanoparticles as MRI contrast agents is increasing due to its short relaxation time (~0.5 ps) and high magnetic moment (10.6 uB), thus giving high contrasts in higher field strength [86].

The incorporation of Dy and other rare earth ions into a matrix for both fluorescent and MR imaging is rarely reported. Dy^{3+} coupled to a noncytotoxic indium phosphide/zinc sulfide (InP/ZnS) QDs that provides fluorescence detection, MR responsiveness, and payload delivery were reported by Grant and coworkers [87]. By surface modifying the QDs with a cell-penetrating peptide sequence coupled to an MR contrast agent, the bimodal nanomaterials functioned as a self-transfecting high-field MR/optical contrast agent for nonspecific intracellular labeling. Fluorescent images confirmed sequestration in perinuclear vesicles of labeled cells, with no apparent toxicity. These techniques could be extended to impart cell selectivity or act as a delivery vehicle for genetic or pharmaceutical interventions. NaDyF4:Tb^{3+} nanocrystals were also investigated as multifunctional negative T_2 MRI contrast agents in high magnetic field and fluorescence imaging [88]. Yet, in another study, Das *et al.* reported single-phased ultrasmall Dy_2O_3:Tb^{3+} nanocrystals as dual modal-contrast agent [89]. Interestingly, the phantom images taken at 7.0 T MRI and *in vivo* MRI images of a mouse model demonstrated that the 4 nm nanocrystals exhibit excellent positive T_2 contrast, which was due to the small r2 relaxivity (2.17 $mM^{-1}s^{-1}$) (Fig. 5.20). Contrast agents showed both T_1 and T_2 relaxation to a different extent. T_1-weighted positive contrast agents showed T_1 and T_2 relaxation, but shortening of T_1 relaxation is dominated over that of T_2 relaxation. This results in bright images. However, species with high T_1 values lend themselves to darker images. Therefore, the resultant smaller value indicated larger T_2-relaxation time, which caused bright contrast. *In vitro* confocal fluorescent microscopy images of the NCs-loaded cells showed that the nanoparticles had been internalized by the cells and emitted green fluorescence. *In vitro* cytotoxicity also indicated that the NCs exhibited minimal toxicity.

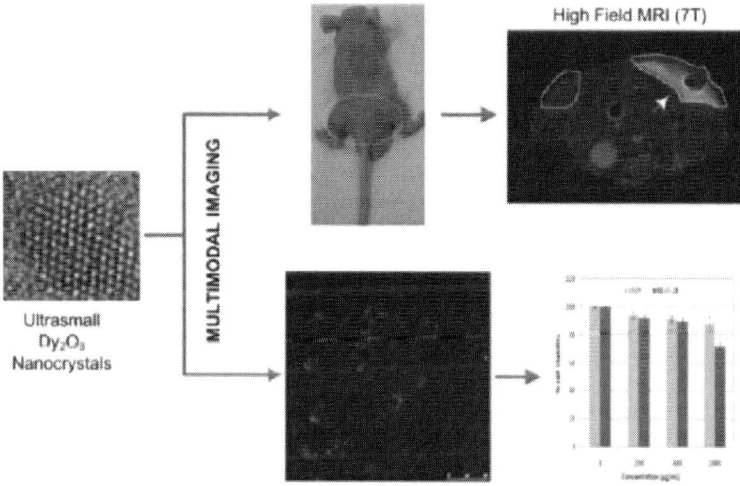

Figure 5.20 The ultrasmall Dy_2O_3 nanocrystals with bright T_2 contrast images in MRI, green fluorescence, and negligible cytotoxicity [89].

The investigation of Dy complex or nanoparticles as contrast agent in multifunctional imaging is fairly unexplored, and it is expected to see continued interest in them in the years to come.

5.2.2 Other Multimodality Contrast Agents and Their Bioapplications

Many recent research efforts have been directed to the investigation of MRI–optical imaging multimodal contrast agents. The fabrication of other multimodality contrast agents is still in their infancy, despite holding great promise in preclinical and clinical use. Only a few studies have been reported about rare earth ions based multimodal imaging nanomaterials.

Gustafsson *et al.* reported gadolinium-based MRI contrast agents and iron oxide based probes that can be labeled with ^{64}Cu for PET imaging [90], which have potential for dual-mode PET/MRI imaging of vascular inflammation. These molecular imaging agents may enable *in vivo* assessment of macrophage accumulation associated with cardiovascular disease. Another example is multimodal image contrast agent for CT and MRI [91]. Gold nanoparticles were functionalized within a multilayered organic shell, which was composed of gadolinium chelates bound to each other through disulfide bonds (Fig. 5.21, left). The contrast enhancement in MRI arose from gadolinium ions (Fig. 5.21, right), and gold core provided a strong X-ray absorption (Fig. 5.22). The study revealed that the particles suited for dual modality imaging could freely circulate in the blood vessels without undesirable accumulation in the lungs, spleen, and liver.

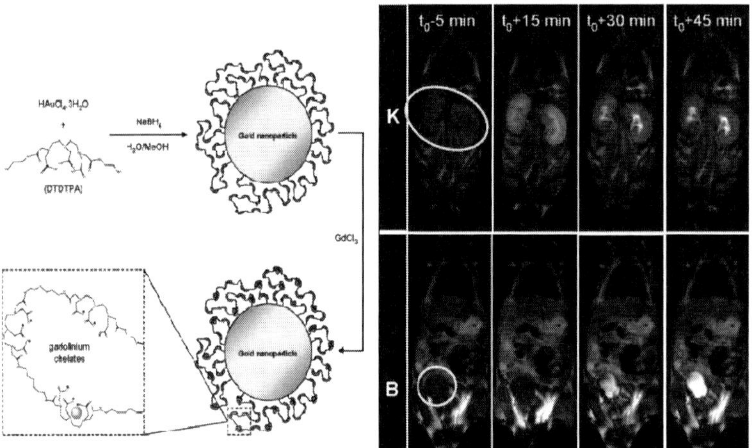

Figure 5.21 Synthesis of Au@DTDTPA-Gd nanoparticles (red circle) Gd^{3+} Ion (left) and T_1-weighted images of a mouse 5 min before (t 0–5 min) and 15, 30, and 45 min after intravenous injection of Au@DTDTPA-Gd_{50} (K for kidneys and B for bladder)(right) [91].

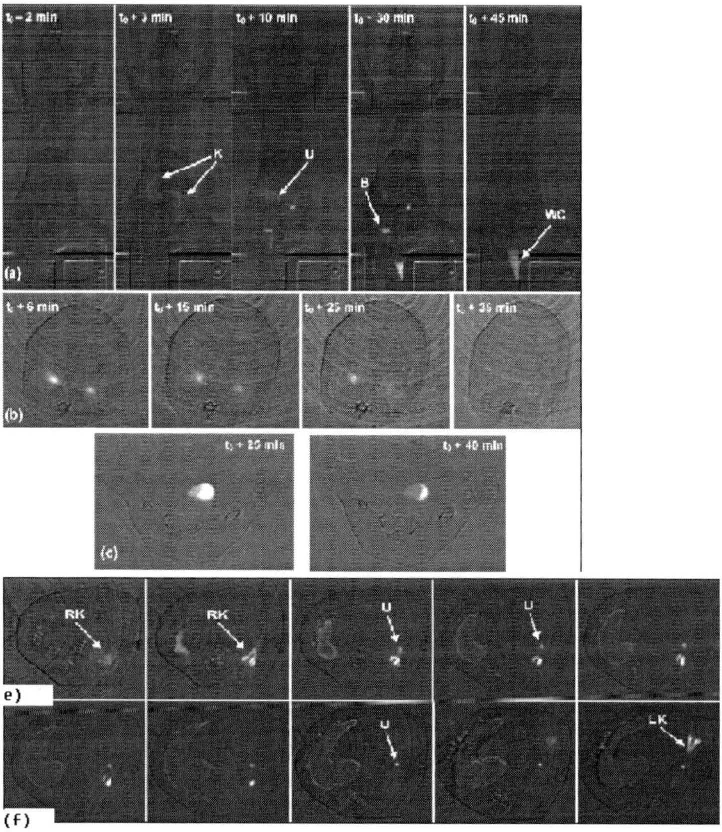

Figure 5.22 (a) Planar X-ray images in transmission mode of a rat before (0–2 min) and after injection of Au@DTDTPA-Gd_{50} (K for kidneys, U for ureter, B for bladder, and WC for the tube collecting the urine). SRCT images of transverse slices recorded at various times after injection of Au@DTDTPA-Gd_{50} to a rat, including (b) kidneys and (c) bladder (slice thickness: 1 mm). SRCT images of a series of successive transverse slices, including right kidney (RK) of a rat from the head (e) to the tail (f). The images were recorded 10 min after the intravenous injection of Au@DTDTPA-Gd_{50} to a rat (RK and LK for right and left kidneys, respectively; U for ureter) [91].

5.3 Conclusions

The fusion of different imaging modalities has unraveled new possibilities for more comprehensive and accurate diagnosis of diseases. Design and fabrication of new multifunctional contrast agents that enable enhanced detection, diagnosis, and imaging at molecular levels for such integrated imaging modality are, therefore, of great importance. Rare earth based nanomaterials, due to their interesting magnetic and fluorescent properties, have attracted strong interests as potential multimodal contrast agents. Most recent studies reported the fabrication of MRI/optical imaging rare earth based contrast agents, and a significant number of such studies demonstrated *in vitro* and, to a lesser extent, *in vivo* MRI and optical imaging. Continual and concerted efforts are needed to translate such application to preclinical or clinical applications. Issues to be addressed include nanomaterials biocompatibility, higher contrast for imaging, and also the presence of a chemically and biologically active surface with a variety of functional groups for conjugation of therapeutic agents and/or for cell targetability. Toxicity is another major issue needed to be addressed. More comprehensive *in vivo* and *in vitro* pharmacological and toxicological studies are needed to thoroughly assess the safety of such nanomaterials.

References

1. Weissleder, R., and Pittet, M. J. (2008) Imaging in the era of molecular oncology, *Nature*, **452**(7187), 580–589.
2. Weissleder, R., and Mahmood, U. (2001) Molecular imaging, *Radiology*, **219**(2), 316–333.
3. Weissleder, R. (2006) Molecular imaging in cancer, *Science*, **312**(5777), 1168–1171.
4. Cassidy, P. J., and Radda, G. K. (2005) Molecular imaging perspectives, *J. R. Soc. Interface*, **2**(3), 133–144.
5. Jaffer, F. A., and Weissleder, R. (2005) Molecular imaging in the clinical arena, *J. Am. Med. Assoc.*, **293**(7), 855–862.
6. Louie, A. (2010) Multimodality imaging probes: design and challenges, *Chem. Rev.*, **110**, 3146–3195.
7. Lakkaraju, A., *et al.* PET/CT in primary musculoskeletal tumours: a step forward, *Eur. Radiol.*, **20**(12), 2959–2972, DOI: 10.1007/s00330-010-1862-z
8. Maurer, A. H. (2008) Combined imaging modalities: PET/CT and SPECT/CT, *Health Phys*, **95**(5), 571–576.

9. Bolus, N. E., *et al.* (2009) PET/MRI: the blended-modality choice of the future? *J. Nucl. Med. Technol.*, **37**(2), 63–71.
10. Weissleder, R., and Rudin, M. (2003) Molecular imaging in drug discovery and development, *Nat. Rev.*, **2,** 123–131.
11. Huh, Y. M., *et al.* (2005) *In vivo* magnetic resonance detection of cancer by using multifunctional magnetic nanocrystals, *J. Am. Chem. Soc.*, **127**(35), 12387–12391.
12. Lee, H., *et al.* (2007) Thermally cross-linked superparamagnetic iron oxide nanoparticles: synthesis and application as a dual imaging probe for cancer *in vivo*, *J. Am. Chem. Soc.*, **129**(42), 12739–12745.
13. Insin, N., *et al.* (2008) Incorporation of iron oxide nanoparticles and quantum dots into silica microspheres, *ACS Nano*, **2**(2), 197–202.
14. Li, L., *et al.* (2007) Magnetic and fluorescent multifunctional chitosan nanoparticles as a smart drug delivery system, *Nanotechnology*, **18**(40), doi: 10.1088/0957-4484/18/40/405102.
15. Zhang, H., Yee, D., and Wang, C. (2008) Quantum dots for cancer diagnosis and therapy: biological and clinical perspectives, *Nanomedicine*, **3**(1), 83–91.
16. Zhang, Y., *et al.* (2008) Self-assembly multifunctional nanocomposites with Fe_3O_4 magnetic core and CdSe/ZnS quantum dots shell, *J. Biomed. Mater. Res. Part A*, **85**(3), 840–846.
17. He, Y., Wang, H. F., and Yan, X. P. (2008) Exploring Mn-doped ZnS quantum dots for the room-temperature phosphorescence detection of enoxacin in biological fluids, *Anal. Chem.*, **80**(10), 3832–3837.
18. Yong, K. (2009) Mn-doped near-infrared quantum dots as multimodal targeted probes for pancreatic cancer imaging, *Nanotechnology*, **20**(1), doi: 10.1088/0957-4484/20/1/015102.
19. Engström, M., *et al.* (2006) High proton relaxivity for gadolinium oxide nanoparticles, *Magn. Reson. Mater. Phys. Biol. Med.*, **19**(4), 180–186.
20. Shen, J., Sun, L. D., and Yan, C. H. (2008) Luminescent rare earth nanomaterials for bioprobe applications, *Dalton Trans.*, **42**, 5687–5697.
21. Hyon Bin Na, I. C. S., and Hyeon, T. (2009) Inorganic nanoparticles for MRI contrast agents, *Adv. Mater.*, **21,** 2133–2148.
22. Basilion, J. P., Yeon, S., and Bodner, R. (2005) Magnetic resonance imaging: utility as a molecular imaging modality, *Curr. Top. Dev. Biol.*, **70,** 1–33.
23. Chatziioannou, A. F. (2002) PET scanners dedicated to molecular imaging of small animal models, *Mol. Imaging Biol.*, **4**(1), 47–63.

24. Dobrucki, L. W., and Sinusas, A. J. (2010) PET and SPECT in cardiovascular molecular imaging, *Nat. Rev. Cardiol.*, **7**(1), 38–47.

25. Mittal, G., *et al.* (2004) Principles and applications of positron emission tomography — a review, *Biomedicine*, **24**(1–2), 45–57.

26. Lapp, R. T., *et al.* Clinical utility of positron emission tomography/ computed tomography in inflammatory bowel disease, *Mol. Imaging Biol.*, **13**(3), 573–576, DOI: 10.1007/s11307-010-0367-0

27. Papathanassiou, D., Bruna-Muraille, C., Liehn, J.-C., Nguyen, T. D., and Curé, H. (2009) Positron emission tomography in oncology: present and future of PET and PET/CT, *Crit. Rev. Oncol./Hematol.*, **72,** 239–254.

28. Miller, P. W., *et al.* (2008) Synthesis of 11C, 18F, 15O, and 13N radiolabels for positron emission tomography, *Angew. Chem. Int. Ed.*, **47**(47), 8998–9033.

29. http://en.wikipedia.org/wiki/Positron_emission_tomography.

30. Madsen, M. T. (2007) Recent advances in SPECT imaging, *J. Nucl. Med.*, **48**(4), 661–673.

31. Mundler, O. (2009) Imagerie hybride, TEP-TDM et TEMP-TDM. En quoi peut-elle influencer la formation et la pratique en médecine nucléaire en France?, *Médecine Nucléaire*, **33**(2), 108–113.

32. Sosnovik, D. E., Nahrendorf, M., and Weissleder, R. (2007) Molecular magnetic resonance imaging in cardiovascular medicine, *Circulation,*. **115**(15), 2076–2086.

33. Gallagher, F. A. (2010) An introduction to functional and molecular imaging with MRI, *Clin. Radiol.*, **65**(7), 557–566.

34. Chan, K. W. Y., and Wong, W. T. (2007) Small molecular gadolinium (III) complexes as MRI contrast agents for diagnostic imaging, *Coord. Chem. Rev.*, **251**(17–20), 2428–2451.

35. Na, H. B., *et al.* (2007) Development of a T1 contrast agent for magnetic resonance imaging using MnO nanoparticles, *Angew. Chem. Int. Ed.*, **46**(28), 5397–5401.

36. Waters, E. A., and Wickline, S. A. (2008) Contrast agents for MRI, *Basic Res. Cardiol.*, **103**(2), 114–121.

37. Sosnovik, D. E., Nahrendorf, M., and Weissleder, R. (2008) Magnetic nanoparticles for MR imaging: agents, techniques and cardiovascular applications. *Basic Res. Cardiol.*, **103**(2), 122–130.

38. Groman, E. V., Bouchard, J. C., Reinhardt, C. P., and Vaccaro, D. E. (2007) Ultrasmall mixed ferrite colloids as multidimensional magnetic

resonance imaging, cell labeling, and cell sorting agents, *Bioconjug. Chem.*, **18**, 1763–1771.

39. Małgorzata Norek, E. K., Zeitler, U., and Peters, J. A. (2008) Tuning of the size of Dy2O3 nanoparticles for optimal performance as an MRI contrast agent, *J. Am. Chem. Soc.*, **130**, 5335–5340.
40. Melanie Bottrill, L. K., and Long, N. J. (2006) Lanthanides in magnetic resonance imaging, *Chem. Soc. Rev.*, **35**, 557–571.
41. Pysz, M. A., Gambhir, S. S., and Willmann, J. K. (2010) Molecular imaging: current status and emerging strategies, *Clin. Radiol.*, **65**(7), 500–516.
42. Grier, D. G. (2003) A revolution in optical manipulation, *Nature*, **424**(6950), 810–816.
43. Pierce, M. C., Javier, D. J., and Richards-Kortum, R. (2008) Optical contrast agents and imaging systems for detection and diagnosis of cancer, *Int. J. Cancer*, **123**(9), 1979–1990.
44. Santra, S., Xu, J., Wang, K., and Tan, W. (2004) Luminescent nanoparticle probes for bioimaging, *J. Nanosci. Nanotechnol.*, **4**, 590–599.
45. Bunzli, J. -C. G., Comby, S., Chauvin, A.-S., and Vandevyver, C. D. B. (2007) New opportunities for lanthanide luminescence, *J. Rare Earths*, **25**(3), 257–274.
46. Smith, A. M., *et al.* (2008) Bioconjugated quantum dots for *in vivo* molecular and cellular imaging, *Adv. Drug Deliv. Rev.*, **60**(11), 1226–1240.
47. Ntziachristos, V., Bremer, C., and Weissleder, R. (2003) Fluorescence imaging with near-infrared light: new technological advances that enable *in vivo* molecular imaging, *Eur. Radiol.*, **13**(1), 195–208.
48. Rieter, W. J., *et al.* (2007) Hybrid silica nanoparticles for multimodal imaging, *Angew. Chem. Int. Ed.*, **46**(20), 3680–3682.
49. Di Carli, M. F. (2009) Hybrid imaging: integration of nuclear imaging and cardiac CT, *Cardiol. Clin.*, **27**(2), 257–263.
50. Gessner, R., and Dayton, P. A. (2010) Advances in molecular imaging with ultrasound, *Mol. Imaging*, **9**(3), 117–127.
51. Cherry, S. R., Louie, A. Y., and Russell E. J. (2008) The integration of positron emission tomography with magnetic resonance imaging. *Proc. IEEE*, **96**(3), 416–438.
52. Jennings, L. E., and Long, N. J. (2009) 'Two is better than one' — probes for dual-modality molecular imaging, *Chem. Commun.*, 24, 3511–3524.

53. Fernández, B., *et al.* (2008) Quantum dots decorated with magnetic bionanoparticles, *Adv. Funct. Mater*, **18**(24), 3931–3935.

54. Jin, T., *et al.* (2008) Gd^{3+}-functionalized near-infrared quantum dots for *in vivo* dual modal (fluorescence/magnetic resonance) imaging, *Chem. Commun.*, **44**, 5764–5766.

55. Gerion, D., *et al.* (2007) Paramagnetic silica-coated nanocrystals as an advanced MRI contrast agent, *J. Phys. Chem. C*, **111**(34), 12542–12551.

56. Prinzen, L., *et al.* (2006) Optical and magnetic resonance imaging of cell death and platelet activation using annexin A5-functionalized quantum dots, *Nano Lett.*, **7**(1), 93–100.

57. Heesun Yang, S. S., Walter, G. A., and Holloway, P. H. (2006) GdIII-functionalized fluorescent quantum dots as multimodal imaging probes, *Adv. Mater.*, **18**(21), 2890–2894.

58. Willem, J. M. M., *et al.* (2006) Quantum dots with a paramagnetic coating as a bimodal molecular imaging probe, *Nano Lett.*, **6**(1), 1–6.

59. Tsai, C.P., *et al.* (2008) High-contrast paramagnetic fluorescent mesoporous silica nanorods as a multifunctional cell-imaging probe, *Small*, **4**(2), 186–191.

60. Hsiao, J.-K., *et al.* (2008) Mesoporous silica nanoparticles as a delivery system of gadolinium for effective human stem cell tracking, *Small*, **4**(9), 1445–1452.

61. Kathryn, M. L., *et al.* (2008) Mesoporous silica nanospheres as highly efficient MRI contrast agents, *J. Am. Chem. Soc.*, **130**, 2154–2155.

62. Kim, J. S., *et al.* (2007) Self-assembled hybrid nanoparticles for cancer-specific multimodal imaging, *J. Am. Chem. Soc.*, **129**(29), 8962–8963.

63. Santra, S., *et al.* (2005) Synthesis and characterization of fluorescent, radio-opaque, and paramagnetic silica nanoparticles for multimodal bioimaging applications, *Adv. Mater.*, **17**, 2165–2169.

64. Li I.-F., Yeh, C.-S. (2010) Synthesis of Gd doped CdSe nanoparticles for potential optical and MR imaging applications, *J. Mater. Chem.*, **20**(11), 2079–2081.

65. Bridot, J.-L. (2007) Hybrid gadolinium oxide nanoparticles: multimodal contrast agents for *in vivo* imaging, *J. Am. Chem. Soc.*, **129**(16), 5076–5084.

66. Petoral, R. M., *et al.* (2009) Synthesis and characterization of Tb^{3+}-doped Gd_2O_3 nanocrystals: a bifunctional material with combined fluorescent labeling and MRI contrast agent properties, *J. Phys. Chem. C*, **113**(17), 6913–6920.

67. Das, G. K., *et al.* (2010) Gadolinium oxide ultranarrow nanorods as multimodal contrast agents for optical and magnetic resonance imaging, *Langmuir*, doi: 10.1021/la904751q.

68. Setua, S., *et al.* (2010) Folate receptor targeted, rare earth oxide nanocrystals for bi-modal fluorescence and magnetic imaging of cancer cells, *Biomaterials*, **31**(4), 714–729.

69. Huang, C.-C., *et al.* (2009) Bifunctional Gd_2O_3/C nanoshells for MR imaging and NIR therapeutic applications, *Adv. Funct. Mater.*, **19**(2), 249–258.

70. Miyawaki, J., *et al.* (2009) Biodistribution and ultrastructural localization of single-walled carbon nanohorns determined *in vivo* with embedded Gd_2O_3 labels, *ACS Nano*, **3**(6), 1399–1406.

71. Das, G. K., *et al.* (2010) Gadolinium oxide ultranarrow nanorods as multimodal contrast agents for optical and magnetic resonance imaging, *Langmuir*, doi:10.1021/la904751q.

72. Stouwdam, J. W, and van Veggel, F. C. J. M. (2002) Near-infrared emission of redispersible Er^{3+}, Nd^{3+}, and Ho^{3+} doped LaF_3 nanoparticles, *Nano Lett.*, **2**, 733–737.

73. Hon-Tung Wong, H. L. W. C., and Hao, J. H. (2009) Magnetic and luminescent properties of multifunctional GdF_3: Eu^{3+} nanoparticles, *Appl. Phys. Lett.*, **95**, 022512.

74. Hon-Tung Wong, H. L. W. C., and Hao, J. (2010) Towards pure near-infrared to near-infrared upconversion of multifunctional GdF_3: Yb^{3+},Tm^{3+} nanoparticles, *Opt. Express*, **18**(6), 6123–6130.

75. Kumar, R., *et al.* (2009) Combined optical and MR bioimaging using rare earth ion doped $NaYF_4$ nanocrystals, *Adv. Funct. Mater.*, **19**, 853–859

76. Kumar, R., Nyk, M., Ohulchanskyy, T. Y., Flask, C. A., and Prasad, P. N. (2009) Combined optical and MR bioimaging using rare earth ion doped $NaYF_4$ nanocrystals, *Adv. Funct. Mater.*, **19**(6), 853–859.

77. Park, Y. I., *et al.* (2009) Nonblinking and nonbleaching upconverting nanoparticles as an optical imaging nanoprobe and T1 magnetic resonance imaging contrast agent, *Adv. Mater.*, **21**(44), 4467–4471.

78. Zhou, J., *et al.* (2010) Dual-modality *in vivo* imaging using rare earth nanocrystals with near-infrared to near-infrared (NIR-to-NIR) upconversion luminescence and magnetic resonance properties, *Biomaterials*, **31**(12), 3287–3295.

79. Li, Z., Zhang, Y., Shuter, B., and Muhammad, I. N. (2009) Hybrid lanthanide nanoparticles with paramagnetic shell coated on upconversion fluorescent nanocrystals, *Langmuir*, **25**(20), 12015–12018.

80. He, H., *et al.* (2009) Synthesis of Fe_3O_4@LaF_3: Ce, Tb nanocomposites with bright fluorescence and strong magnetism, *Appl. Surf. Sci.*, **255**(8), 4623–4626.

81. Zhang, Y., *et al.* (2009) Tb-doped iron oxide: bifunctional fluorescent and magnetic nanocrystals, *J. Mater. Chem.*, **19**(22), 3696–3703.

82. Lu, H., *et al.*, (2004) Synthesis and characterization of multi-functional nanoparticles possessing magnetic, up-conversion fluorescence and bio-affinity properties, *J. Mater. Chem.*, **14**(8), 1336–1341.

83. van Tilborg, G. A. F., *et al.* (2006) Annexin A5-functionalized bimodal lipid-based contrast agents for the detection of apoptosis, *Bioconjug. Chem.*, **17**(3), 741–749.

84. Lu, H., *et al.* (2004) Synthesis and characterization of multi-functional nanoparticles possessing magnetic, up-conversion fluorescence and bio-affinity properties, *J. Mater. Chem.*, **14**, 1336–1341.

85. Bottril M., Kwok L., and Long J. N., (2006) Lanthanides in magnetic resonance imaging, *Chem. Soc. Rev.*, **35**, 557–571.

86. Viswanathan S., Kovacs Z., Green K. N., Ratnakar S. J., and Sherry A. D. (2010) Alternatives to Gadolinium-based metal chelates for magnetic resonance imaging, *Chem. Rev.*, **110**, 2960–3018.

87. Rosenberg J. T., Kogot J. .M., Lovingood D. D., Strouse G. F., and Grant S. C. (2010) Intracellular bimodal nanoparticles based on quantum dots for high-field MRI at 21.1 T, *Magn. Reson. Med.*, **64**(3), 871–882.

88. Zhang Y., Vigayaragavan V., Das G. K., Bhakoo K. K., and Tan T. T. Y., Single-phase $NaDyF_4$:Tb nanocrystals as multifunctional contrast agent in high field magnetic resonance and optical imaging, *Eur. J. Inorg. Chem.* (in press).

89. Das G. K., Zhang Y., D' Silva L., Padmanabhan P., Heng B. C., Loo J. S. C., Selvan S. T., Bhakoo K. K., and Tan T. T. Y. (2011) Single-phase Dy_2O_3:Tb^{3+} nanocrystals as dual-modal contrast agent for high field magnetic resonance and optical imaging, *Chem. Mater.*, **23**, 2439–2446.

90. Gustafsson, B., Youens, S., and Louie, A. Y. (2006) Development of contrast agents targeted to macrophage scavenger receptors for MRI of vascular inflammation, *Bioconjug. Chem.*, **17**(2), 538–547.

91. Alric, C., *et al.* (2008) Gadolinium chelate coated gold nanoparticles as contrast agents for both X-ray computed tomography and magnetic resonance imaging, *J. Am. Chem. Soc.*, **130**(18), 5908–5915.

Chapter 6

Rare Earth Nanophosphors in Light-Emitting Diodes

Chin Yun Tee, Gautom Kumar Das, Yan Zhang, and Timothy T.Y. Tan
School of Chemical and Biomedical Engineering, Nanyang Technological University, 637459 Singapore
tytan@ntu.edu.sg

Rare earth doped materials are vital in many optoelectronics applications such as solid-state lasers [1], light-emitting diodes (LEDs) [2], optical fiber amplifiers [3], and photovoltaic devices [4]. Majority of earlier works focus on the rare earth doped materials of bulk nature. However, with the development of nanotechnology over the past decades, research and applications of rare earth materials in optoelectronics have moved from bulk to nanoscale. In this chapter, recent developments of rare earth nanophosphors in LEDs, and the effects of various rare earth dopants and particle size on luminescence properties are discussed.

6.1 Introduction and Overview

Over the past few decades, rare earth materials have been widely studied due to their unique luminescent properties arisen from

Rare Earth Nanotechnology
Edited by Timothy T.Y. Tan

ISBN 978-981-4316-30-9 (Hardcover), 978-981-4364-20-1 (eBook)
www.panstanford.com

their sharp emission bandwidths, tunable emission color, long luminescent lifetime, and low toxicity. RE ions exhibit up- and down-conversion luminescence [5]. Up-conversion is a process whereby photons are pumped by a longer wavelength excitation, which produces emission of shorter wavelength. This is applicable in the field of biomedical research where the up-conversion processes of RE doped materials proved to be promising candidates for biological assays and biomedical imaging [6–9]. However, the down-conversion process, the reverse of up-conversion, is the typical mechanism of light emission in LEDs [10–12].

Rare earth phosphors are under extensive investigation during the past decades due to the increasing demand of higher energy efficient light sources and luminescent panels [13]. It has driven developments targeted to create optoelectronic devices such as white light emitting diodes [2, 14, 15] and display panels with enhanced efficiency [16, 17]. Traditional light bulbs, which are widely used in incandescent lamps, are very energy inefficient, where only 3–5% of electrical energy is converted into visible light. Conventional incandescent lamps rely on either heat or discharge of gases. Both processes involve large energy losses due to high temperatures and large Stokes shifts. In order to make energy efficient systems, efforts have been made to replace conventional incandescent lamps with more efficient white LEDs, which reduce energy consumption and greenhouse gas emissions [18].

The investigation of rare earth materials in optoelectronics started in the beginning of the 20th century with the emphasis on bulk material [19–21]. Due to higher surface area-to-volume ratio, nanomaterials are usually more reactive and sensitive than their bulk counterparts. In order to enhance efficiency, most recent development in rare earth doped phosphors for optoelectronics applications are moving toward nanosize particle [22, 23].

Various rare earth based nanophosphors used in white LED, primarily yellow, red, green, and blue emitting phosphors will be discussed. This chapter will begin with a brief explanation on the working principles of LED and how white light is produced in LEDs using rare earth doped nanomaterials. This is followed by an overview of the various synthesis methods of the nanophosphors and key factors affecting luminescence efficiency. The bulk of the chapter will concentrate on the effect of dopant, structure, host materials and particle size on the luminescence properties of the various rare earth nanophosphors.

6.2 How Does LED Work?

Similar to a regular diode, LED is a device made of semiconducting material doped with impurities to create a p-n junction. Light is emitted when an electric current is passed through the semiconductor device. When the p-n junction in the LED is under forward biased, the holes in the p-type material are pushed toward the junction while the electrons in the n-type material are pushed toward the junction (Fig. 6.1). This creates a narrower depletion region and lowers the potential barrier. With continual forward-biased voltage, the electrical resistance at the junction may be overcome. The electrons and holes combined at the junction, maintaining a continuous current. Electrons from the conduction band flow from the n-type region to combine with the holes from the valence band from the p-type region. As the electron and hole recombine at the p-n junction, light with an energy $h\nu = E_g$ is emitted. Wavelength of light emitted is corresponding to the bandgap energy of the semiconductor material [24].

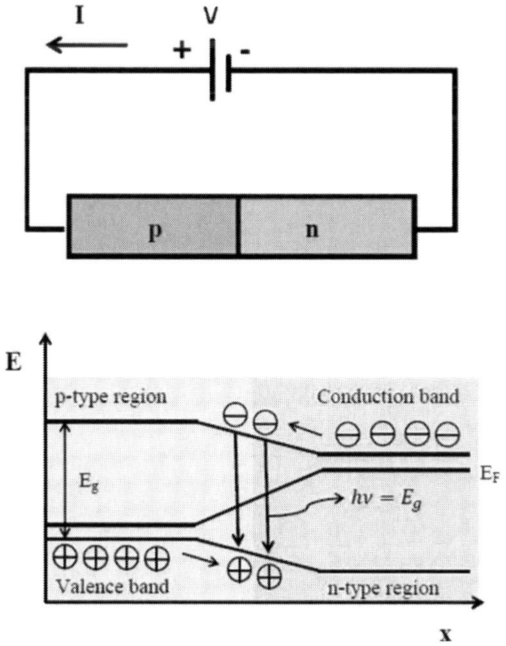

Figure 6.1 Potential energy vs distance graph showing the movement of holes and electrons at the p-n junction [25].

All commercially available LEDs in the visible to the infrared range (i.e., 450–1500 nm) are made of semiconductor thin films of compounds consisting of elements from group three and five in the periodic table (III–V semiconductors) [26]. Group three elements include aluminum, gallium, and indium, while group five elements include nitrogen, phosphorus, and arsenic. Common compounds include GaN and InGaN.

As shown in Fig. 6.2, LED is typically made up of three layers of different materials [24, 27]. The base layer is an n-type material (e.g., GaN doped with Si) with a high concentration of free electrons, followed by multiple alternating thin layers of materials with smaller bandgap known as quantum wells. The top layer is a p-type material (e.g., GaN doped with Mg) with a high concentration of holes. The region with smaller bandgap sandwiched between the two layers of larger bandgap materials creates a region where the holes and electrons may recombine, emitting light with the wavelength of the smaller bandgap material.

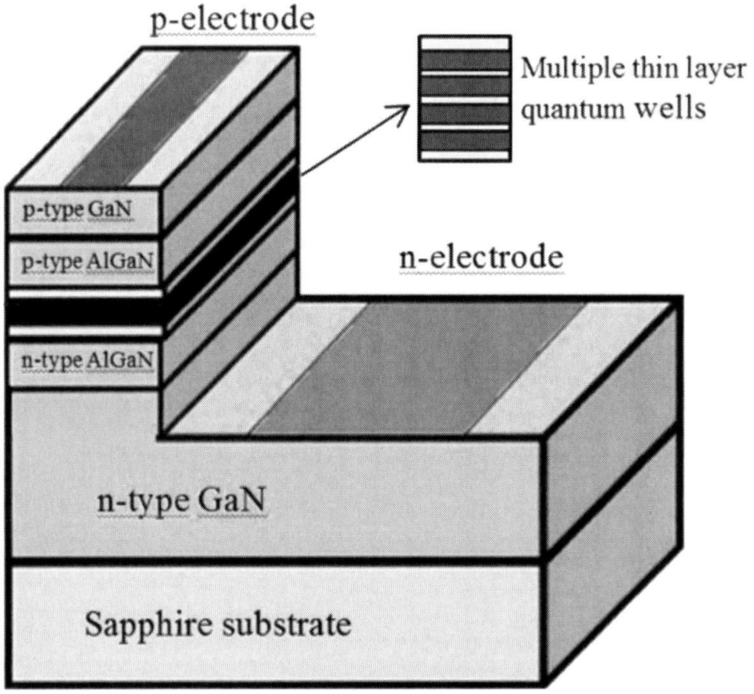

Figure 6.2 Structure of a typical GaN-based LED [25].

6.2.1 Production of White Light from LED

There are three dominant methods of producing high intensity white light using LEDs (Fig. 6.3). The first method is by using yellow-, green-, or red-emitting phosphors to convert monochromatic light from blue LED to white light. The second method is by using blue-, green-, and red-emitting phosphors to convert monochromatic light from ultraviolet (UV) light LED to white light. The third method is by using three individual monochromatic blue, green, and red LEDs (Fig. 6.4) [18]. The three different schemes are adopted for different applications as all three have their advantages and disadvantages. The first method where blue LED is used with a yellow-emitting phosphor is the most common. It has a high theoretical efficacy, which makes it attractive as a cheap, bright white light source. However, it has a low color rendering index (CRI) value. CRI refers to the ability to reproduce colors of an object as seen under an ideal white light source, such as the sun. A low CRI would mean that the white LED is unsuitable for indoor use as the intensity of white light from the LED is insufficient to illuminate an object. The second method, which produces UV LEDs, has low theoretical efficacy but higher CRI values; thus it is more suitable for indoor use. The third method has a higher efficacy than UV LEDs; however, it is by far the most expensive method of producing white LED [24]. In this chapter, the third method will not be discussed. Our focus will be mostly limited to rare earth doped phosphors for white LEDs.

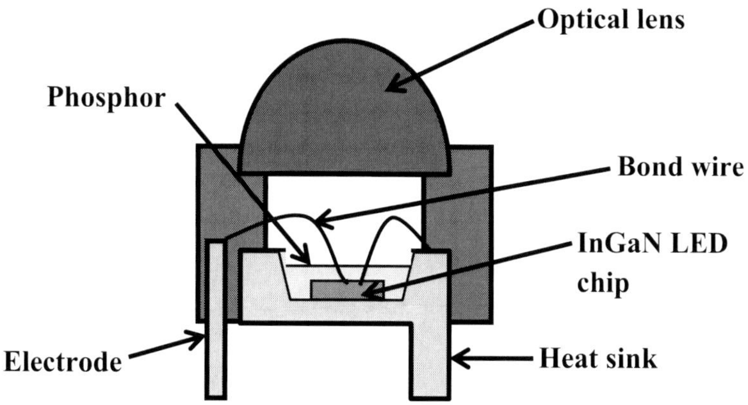

Figure 6.3 Structure of white LED [28].

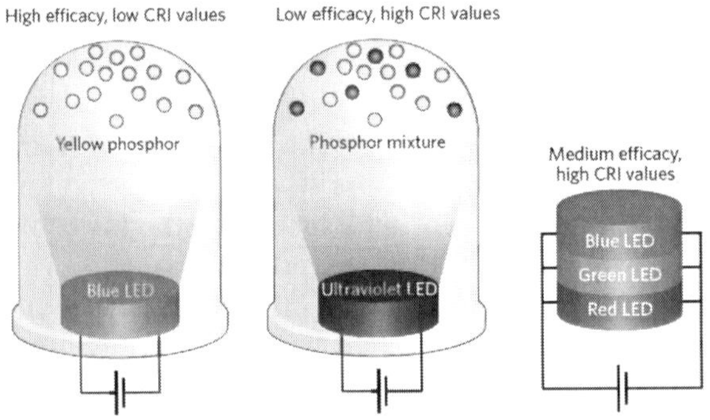

Figure 6.4 The different methods of producing white light based on LEDs [24].

Rare earth phosphors are widely chosen for white LED applications due to the unique luminescence properties of rare earth ions in contrast to other luminescent centers adopted such as transition metals and complex anions. Broadband emissions are observed in the partially filled d shells of transition metal ions due to d-d transitions. However, rare earth ions have sharp emission bands due to f-f transitions. This two transitions, d-d transitions and f-f transitions, are parity and spin forbidden, hence are slow [118]. For rare earth ions, broadband emissions due to 4f-5d transitions are achievable as rare earth ions have 4f and 5d shells. This f-d transition gives broadband emission and faster transition rates, which are more favorable for phosphor applications. Further, the spectral position of emission lines is independent on host lattice in rare earth ions, in contrast to the emission lines generated by transition metals, which results in chemical bonding change during transition. Chemical bonding change is also exhibited in complex anions like $CaWO_4$ during charge transfer transition [119]. Faster f-d transition rates in rare earth ions contribute to higher luminescence efficacies without significant chemical bonding change, which makes rare earth phosphor a good choice for light emitting applications.

6.3 Synthesis of Rare Earth Doped Phosphors

The synthetic routes of rare earth phosphors can be categorized into two main groups: the "dry" methods and the "wet" methods. The "dry" methods refer to synthetic routes like the solid-state reaction, self-propagating synthesis, and several other physical routes like chemical vapor deposition (CVD) and magnetic sputtering. These dry methods commonly require rigorous conditions such as high reaction temperature and extreme conditions, hence may not be perceived as favorable [29].

Conventional bulk YAG:RE phosphors are synthesized based on the solid-state reaction method. However, in order to achieve phosphors with large degree of homogeneity and smaller particle size, wet methods synthetic routes have been adopted [30]. The wet methods generally involve solution-based processes, where various parameters may be varied, and thus phase purity, chemical homogeneity, particle size, and morphology can be controlled. In contrast to the dry methods, the wet methods are usually carried out under milder reaction conditions and are more energy efficient and economical. Rare earth doped nanophosphors are mainly synthesized based on wet chemical methods.

In the following sections, solid-state synthesis for bulk phosphors will be discussed first. Other synthesis techniques such as coprecipitation [31–34], sol gel [35–39], and hydrothermal/solvothermal methods [40–43] will also be discussed. These techniques are applied for synthesis of both rare earth nanosize phosphors as well as phosphors of smaller size as compared to bulk phosphors, hence more superior in luminescence properties. For more discussion on rare earth nanomaterials synthesis, please refer to Chapter 1.

6.3.1 Solid-State Synthesis of Bulk YAG:Ce Phosphors

The solid-state synthesis method conventionally involves high temperatures where the starting materials are first mixed thoroughly with fluxes. Fluxes are substances added to raw material to help crystal grow. After that, the mixture is calcined in a crucible at high temperature and under reduced temperature for a few hours. For example, starting materials for YAG:Ce phosphor synthesis under this method are Y_2O_3, CeO_2, and Al_2O_3, weighed according to stoichiometric ratio of the chemical formula $(Y_{2.95}Ce_{0.05})Al_5O_{12}$. These starting materials are mixed thoroughly with fluxes such as H_3BO_3, BaF_2 or alkali fluorides. Then, the mixture is calcined under a reductive atmosphere (95% N_2/5% H_2), at 1500°C for 6 h [44, 45]. Shao *et al.* reported the synthesis of Gd^{3+} and La^{3+} doped YAG:Ce phosphor, which used the same starting material as above with the addition of La_2O_3 and Gd_2O_3. The calcinating temperature and reductive pressure also differed slightly at 1550°C for 6 h and 25% N_2/75% H_2, respectively [46]. The solid-state reaction method may not be favored due to high temperatures, agglomeration, and impurities introduced from the fluxes [47] .

6.3.2 Co-Precipitation Synthesis of Rare Earth Phosphors

The co-precipitation route yields pure products at relatively low temperatures and has relatively simple steps. Typically, metal ions such as nitrates and chlorides are first precipitated from solutions such as carbonates, oxalates, and hydroxides. Following this, the precipitate is filtered, washed, and dried. After that, the dried precipitate undergoes calcination treatment to obtain the final products.

Chiang *et al.* [33] reported the synthesis of Ce^{3+} doped terbium aluminum garnet (TAG) phosphor using cerium, terbium, and aluminum nitrate salt compounds, dilute in HNO_3 solution to prepare 250 ml stoichiometric ratio of mixing solution. NH_4HCO_3 was then titrated with the solution for precipitation. The precursor precipitate was filtered, washed four times with de-ionized water, and dried at 100°C for 12 h. The precursor was calcined at temperature between 500–850°C for 2 h under reduction atmosphere of CO to obtain products. The precursor was amorphous when calcined at temperatures below 850°C. Pure TAG:Ce phosphor was formed when calcined at 1000°C. Yu-ling *et al.* [34] also reported a similar method for the synthesis of $(Ca_{1-x-y}, Ln_y)MoO_4$: xEu^{3+} (Ln = Y, Gd) red phosphors. All the reaction conditions and steps are the same as above, the only difference being the starting reagents.

6.3.3 Sol-Gel Synthesis of LED Rare Earth Phosphors

The sol-gel method is commonly used as it has distinct advantages over the solid-state reaction method. For instance, the starting materials have higher reactivity and molecular mixing of constituents, which lead to chemical homogeneity and shorter diffusion distances for the reactants, hence requiring lower calcinations temperatures [35, 36, 39].

Fadlalla *et al.* and Rongfeng *et al.* demonstrated synthesis of YAG:Ce phosphor via the sol-gel route [35, 37]. The main starting materials were yttrium nitrate, $Y(NO_3)_3{\cdot}6H_2O$, aluminum potassium sulphate dodecahydrate, $KAl(SO_4)_2{\cdot}6H_2O$, cerium nitrate, $Ce(NO_3)_3{\cdot}6H_2O$. The aluminum source may be replaced by aluminum nitrate, $Al(NO_3)_3{\cdot}9H_2O$ [30, 35]. A chelating agent like citric acid, ($C_6H_8O_7{\cdot}H_2O$), and polymerization agent like ethylene glycol (EG) were required. The general steps were as follows: $Y(NO_3)_3{\cdot}6H_2O$ aqueous solution was added drop-wise to a beaker containing the aluminum aqueous solution. The mixture was added into $Ce(NO_3)_3{\cdot}6H_2O$ aqueous solution. The whole solution was stirred at room temperature for 20 min. The above solution was then added dropwise to a beaker containing citric acid dissolved in EG with constant magnetic stirring until a transparent mixture was obtained. This stirring step was done in a water bath of 60–80°C. The solution was then heated to around 120 or 180°C in an electric oven for 24 h. The result was a brown gel substance, which under further heating at 400°C or 600°C for 2 h transformed into dark solid mass, an amorphous gel. After cooling to room temperature, the solid was grounded into a fine powder. The powder then underwent calcination at 700–1200°C for 5 h in air to give the final products.

6.3.4 Hydrothermal/Solvothermal Synthesis of LED Rare Earth Phosphors

The hydrothermal/solvothermal synthesis has relatively simple steps. The reaction is typically done in autoclaves, which results in superheated solvents and high pressures such that the microenvironment of the experiment is distinctly different from other common synthetic systems. When chelating agents are employed, the final products usually have few defects with high crystallinity and fine homogeneity. Typically, the starting materials were mixed and then heated in autoclave for a few hours at a particular temperature. After the reaction was complete, the precipitate was collected and washed with de-ionized water and ethanol for a few times. Finally, the precipitate was dried to obtain the final products [40–43].

Xiao *et al.* reported the hydrothermal synthesis of Sm^{3+} doped $CaWO_4$ red-emitting phosphor with good luminescence properties promising for white LED applications. $Ca(NO_3)_2{\cdot}4H_2O$ (0.2 M) and $Sm(NO_3)_3$ were mixed with Na_2WO_4 (0.2 M) solution and stirred for 30 min. After which, the mixture was poured into an autoclave, sealed, and heated at 150°C for 6 h. After the reaction, the precipitate was left to cool naturally and then collected, washed, and dried at

100°C for 12 h. The average particle size was between 12 nm and 16 nm depending on Sm concentration [40].

Ding *et al.* also synthesized $SrSi_2N_2O_2$:Yb^{2+}, Nd^{3+} phosphors using the hydrothermal method. Starting raw materials were AIN, $Sr(NO_3)_2$, Na_2SiO_3, $Yb(NO_3)_3$, and $Nd(NO_3)_3$, mixed and stirred. The mixture was then poured into an autoclave and heated to a soaking temperature at 150°C with a heating rate of 10°C/min, followed by cooling, washing, and drying. After the dried products were obtained, the products were fired for 2–6 times at 800–1200°C under reduced atmosphere of N_2-H_2 (2.5–7.5%) in a tube furnace. Analysis showed that the phosphor sintered at 1000°C for 4 h showed the highest PL intensity and is a good host material for yellow-green-emitting phosphor, which generates white light when combined with blue LED [42]. The comparison of various rare earth phosphor synthesis methods and their advantages/disadvantages are summarized in Table 6.1.

6.4 Factors Affecting the Luminescence Properties of Rare Earth Phosphors

In order to improve the luminescence properties of the phosphors, the efficiencies of energy transfer among the dopant ions should be enhanced, while non-radiative recombination should be minimized [48]. These factors are influenced by the energy transfer and excitation mechanisms of the host materials. Godlewski [48] reported that in order to achieve narrow spectra emissions, phosphors were usually doped with transition metals (TM) or rare earth elements. Transition metals doping will not be discussed here.

Rare earth ions can be excited through various mechanisms, such as successive absorption of the pumped photons by a single ion or transferring the absorbed energy to the neighboring ions. For energy transfer, an energy overlap between the energy donor and the energy acceptor is required. In case of rare earth doped phosphors, rare earth ion acts as the energy acceptor. However, the excitation mechanism in rare earth doped mechanism also has its limitations. The 4f-4f intra-shell transitions in rare earth ions are inefficient due to the parity and spin selection rules [49, 50]. Since 4f-4f excitation is hard to take place in rare earth doped materials, 4f-5d charge transfer or excitonic mechanism must be used. [51] In another study, Yatsenenko *et al.* [52] reported efficient luminescence

Table 6.1 Comparison of rare earth phosphors synthesis methods and their advantages and/or disadvantages

Synthesis method (starting materials)	General steps	Reaction conditions (temp/pressure)	Pros and/or cons	Size of particles
Solid State Reaction (Y_2O_3, CeO_2 and Al_2O_3 with flux such as H_3BO_3, BaF_2 or alkali fluorides)	• Thorough mixing of starting materials and flux. • Calcination in a crucible with lid.	Calcination at around 1500°C for 6 h at reductive pressure of $95\%N_2/5\%H_2$	Unfavored due to the high temperatures, agglomeration and impurities introduced due to the presence of fluxes.	Bulk particles.
Coprecipitation (Solution of a mixture of nitrate salt compounds, titrated with carbonates, oxalates or hydroxides.)	• Metal ions such as nitrates and chlorides are first precipitated from solutions such as carbonates, oxalates and hydroxides. • The precipitate is filtered, washed and dried.	Precursor precipitate dried at 100°C for 12 h.	Able to obtain pure phase products at relatively low temperatures and simple steps.	Nanosize or smaller sized particles (compared with bulk particles). Smaller particles with enhanced luminescence properties due to reduced energy loss with smaller surface area to volume ratio and increased luminescence intensity. (Further discussed in section 4 and 5)

(Cont'd)

Table 6.1 *(Cont'd)*

Synthesis method (starting materials)	General steps	Reaction conditions (temp/pressure)	Pros and/or cons	Size of particles
	• Dried precipitates undergo calcination treatment	Calcination at 500–850°C for 2 h under reduction atmosphere of CO. Amorphous precursor calcined at below 850°C. Pure phosphor calcined at 1000°C		
Sol-gel ($Y(NO_3)_3$, $KAl(SO_4)_2$, $Ce(NO_3)_3\cdot$ with a chelating agent like citric acid, $C_6H_8O_7\cdot H_2O$ and polymerisation agent, Ethylene Glycol (EG) .)	• Multiple dropwise addition and mixing of materials. • Stirring in water bath, followed by heating in an oven.	Water bath of 60–80°C. Solution heated at around 120 or 180°C in an electric oven for 24 h.	Starting materials has higher reactivity, molecular mixing of constituents leads to chemical homogeneity and shorter diffusion distances for the reactants led to lower calcinations temperatures. Steps quite tedious.	
	• Further heating to obtain amorphous gel.	Heating at 400°C or 600°C for 2 h.		

Synthesis method (starting materials)	General steps	Reaction conditions (temp/pressure)	Pros and/or cons	Size of particles
	• Cooling to room temperature, then ground to fine powder. • Powder calcined in furnace.	Calcinations at 700–1200°C for 5 h in air to give the starting samples.		
Hydro/solvothermal (Nitrate salt compounds.)	• Starting materials mixed, then heat in autoclave for a few hours. • Collect the precipitate and wash with deionized water and ethanol for a few times. • Dry the precipitate.	After mixing, stirred for around 30 min. Heated in autoclave at 150°C for 6 h. Cool naturally, then collected, washed and dried at 100°C for 12 h.	Relatively simple steps. The reaction is typically done in autoclaves, which results in superheated solvents and high pressures such that the microenvironment of the experiment is distinctly different from other common synthetic systems. The final products usually have few defects with high crystallinity and fine homogeneity.	

of terbium doped ZnMnS nanoparticles due to the 4f-4f transition of the Tb^{3+} ion, via Mn^{2+} ions.

Another energy transfer process that enhances efficiency of rare earth phosphors is the cascade transition process [48]. Currently, the phosphors used in fluorescence lamps seem fairly efficient in converting the UV photons emitted by mercury (Hg) vapor into visible light. This conversion reported very high quantum efficiency as one UV photon emitted by Hg vapor is converted to exactly one visible photon, which relates to 100% conversion efficiency. However, if the conversion is considered in energy terms, the efficiency is much lower. This efficiency of fluorescence phosphors can be improved if one UV photon can excite two or more visible photons in a cascade transition process. This process has been found to exist in various host systems doped with Praseodymium (Pr^{3+}) ions.

The energy structure of Pr^{3+} ions consists of ^3H, ^3F, ^1D, ^1G, ^3P, ^1I, and ^1S multiplets split by spin–orbit interaction, as shown in Fig. 6.5.

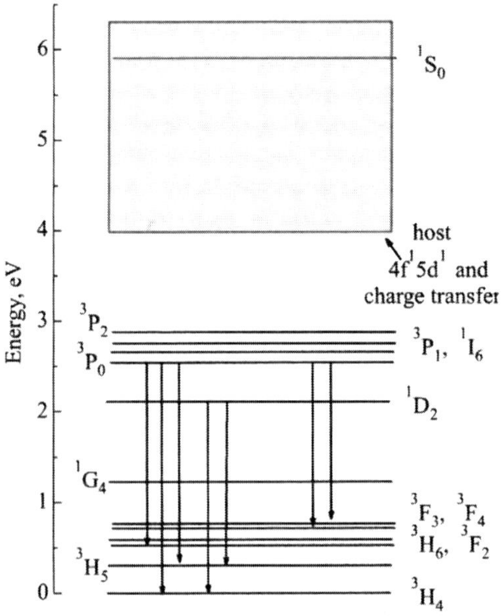

Figure 6.5 The energy level scheme of Pr^{3+} ion in the ZrO_2 lattice with the indicated scheme of cascade processes following the initial 4f-5d charge transfer or host excitation [48].

The photon cascade process observed shows an efficient $4f^2$ $-4f^15d^1$ excitation, followed by carrier trapping via the highest lying 4f state in Pr^{3+}, $^1S_0$. The radiative decay from this $^1S_0$ level then proceeds in a photon cascade process, with the emission of two or more photons: $^1S_0$-$^3P_1$, $^1I_6$; $^3P_0$-$^3H_4$. However, this situation is unfavorable in some host systems like YF_3 and LaF_3 due to two main reasons. First, the first step in the cascade process resulted in UV emission. Secondly, there was evidence of direct recombination from the $^1S_0$ state to the ground ^3H states [53]. These transitions are not useful and should be avoided.

Besides rare earth doping being an important factor in improving luminescence efficiency of phosphors, the reduction of size of phosphors is also important. This is proven by the use of nanoparticles as host lattices in phosphors that showed remarkable luminescence such as in recent studies on Pr^{3+} ions in ZrO_2 nanoparticles [54]. Furthermore, rare earth doped materials in the nanosize range were found to enhance the rate of radiative decay in phosphors as reported by Dicke *et al.* [55]. It was hypothesized that nanoparticles are involved in a process known as superradiance, which refers to the appearance of a fast component of PL decay accompanied by a PL enhancement. This is expected only in the initial stages of PL, when there is a high concentration on excited ions. After which, normal PL decay is observed. Rare earth ions emit at resonant energy when heavily excited, exhibiting superradiance.

In a study, Eu^{3+} doped nanoparticles with two different dopant concentrations (5% and 15%) were investigated. The ultrafast decay process was observed under the same excitation condition. Results indicated that the ultrafast components of the PL decay were identical in both systems, which was most likely heavily dependent on the radiative PL decay in the nanosize phosphor, possibly superradiance [56].

From the discussion above, it is clear that both rare earth doping and the reduction of particle size are important factors influencing the luminescence efficiency in phosphors. Therefore, in the next section where rare earth doped phosphors in white LEDs will be reviewed, the effects of particle size and rare earth dopants on the luminescence characteristics will be emphasized.

In the next sections, recent works on various types of rare earth doped phosphors for white LED applications will be discussed. These phosphors include yellow-, red-, green-, and blue-emitting phosphors. The luminescence property of phosphors is the main factor for determining their applications in LED. Luminescence properties of phosphors are strongly dependent on crystalline properties and particle size, surface morphology, host lattice, as well as type, and concentration of dopant ions [30]. The following discussion will focus mainly on the effects of particle size (where applicable) and crystalline structure, type of rare earth dopant, and dopant concentration. Thermal quenching effects will also be discussed for some phosphors.

6.4.1 Yellow-Emitting Phosphors

Yellow-emitting phosphor combined with blue LED to produce white light is the most commonly adopted method for white LED applications due to lower fabrication cost and higher luminous efficiency. The most popular yellow-emitting phosphor used since the first generation commercial white LED was cerium doped yttrium aluminum garnet phosphor (YAG:Ce). However, YAG:Ce based phosphors have red spectral deficiency and thus have low color rendering properties. As such, other host materials for yellow-emitting phosphors with better light harvesting potential have been investigated [56].

6.4.1.1 Cerium-doped yttrium aluminum garnet phosphors (YAG:Ce)

Rare earth doped yttrium aluminum garnet phosphors (YAG:RE) are important in many luminescence and optical devices. Particularly, cerium-doped YAG:Ce has been widely used in solid-state materials and many lighting applications, especially in white LEDs. This is due to YAG having high chemical stability, high electrical conductivity [30], long lifetime, and resistance to humidity [31].

Wu *et al.* investigated phase development of YAG:Ce powders using X-Ray diffraction (XRD) [57]. The XRD pattern showed that the YAG:Ce particles remained amorphous when maintained at 200°C for 1 h. After 5 h at 200°C, the particles were observed to become pure crystallites. With a longer holding time of up to 10 h, it was found that there were no other phases. While this indicates that the growth of YAG:Ce crystallites was dependent on the length of holding time, the phase purity achieved was desirable for useful applications.

6.4.1.1.1 *Luminescence characteristics of YAG:Ce phosphors*

(a) Particle size and luminescence properties

Wu *et al.* reported YAG:Ce particles synthesized with a holding time of 5 h to be between 200–260 nm and spherical in shape [57]. At this stage, the particles were segregated from each other, which was beneficial for LED phosphors applications.

The excitation bands for YAG:Ce samples were located at 343 nm and 460 nm. These excitation bands were due to the electron transition of Ce^{3+} from the ground state to the excited 5d state with different crystal field splitting components [58]. Ce^{3+} has a $4f^1$ electronic configuration, which can be excited to the $4f^0 5d^1$ state. The 5d electron in the excited state undergoes ^2D term splitting into $^2D_{3/2}$ and $^2D_{5/2}$ states. The 4f electron at the ground $4f^1$ configuration also splits by spin-orbit coupling into $^2F_{7/2}$ and $^2F_{5/2}$ states. In the YAG structure, the lowest $^2D_{3/2}$ state corresponded to the energy of blue light, which gave an excitation band at 460 nm. The emission spectrum of the sample, which has undergone holding time of 5 h, had a maximum peak at 525 nm.

With prolonged heating time, the particle size increased and both excitation and emission intensities increased. This increase in intensity was due to the enhancement of crystallinity and decrease in surface defects [59]. While a larger particle size resulted in a slight red shift in the emission peak, the excitation peak remained largely unchanged (Fig. 6.6). However, there is a greater attraction force between larger particles, which results in aggregation. Hence, smaller particles in YAG:Ce phosphors are favored.

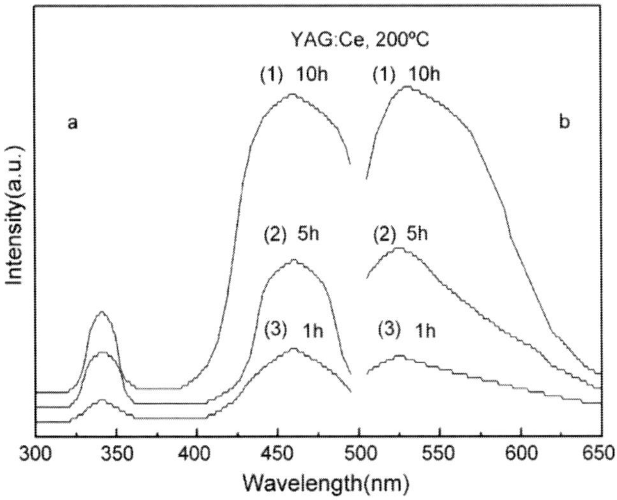

Figure 6.6 Excitation (a) and emission (b) spectra of the sample for different holding time [57].

In Huang *et al.*'s work, it was also found that the smallest YAG:Ce particle size required the least amount to achieve the same percentage intensity of converted visible light in white LED [44]. This suggested that a reduction in particle size would require a minimum amount of phosphor sample without much loss in efficiency in white LED as compared to the amount needed in bulk commercial phosphor. The result of this study is worth noting as it implies that YAG:Ce phosphors with smaller particle size are more cost effective as a smaller amount is needed for white LED.

(b) Effect of dopants on luminescence properties

Yang *et al.* synthesized nanocrystalline YAG:Ce phosphors using a soft chemistry method similar to the sol-gel method. The same procedure was also used to synthesize Pr-doped and Ga/Gd-substituted YAG:Ce nanocrystalline samples. The average size of the sample annealed at 1000°C was about 45 nm and spherical in shape [56].

The PL characteristics of the YAG:Ce sample were similar to that described in the previous section where the two excitation bands involved electronic transitions from the ground state of Ce^{3+} to the crystal field splitting bands of the excited 5d state of the Ce^{3+} ion. Figure 6.7(a) shows the PL emission spectra of YAG:Ce and Pr co-doped YAG:Ce samples using a 440 nm excitation. A clear emission peak at 520 nm can be seen for both samples. However, when doped with red emitting Pr^{3+} ion, an additional red line emission at 608 nm could be seen from the PL emission spectra. This extra line resulted from the energy transfer from the excited Ce^{3+} ion to its neighboring Pr^{3+} ion. Figure 6.7(b) shows the PL excitation spectra of the two samples. The YAG:Ce, Pr sample showed three excitation bands where the first band at 287 nm corresponds to Pr^{3+} excitation, and the other two bands were from Ce^{3+}. The presence of the first additional band showed that there was an energy transfer process from Pr^{3+} to Ce^{3+}, since the 287 nm excitation wavelength could not excite the Ce^{3+} ion directly. This energy transfer process is depicted schematically in Fig. 6.7(c) [60]. Basically, there may be a nonradiative relaxation from the excited $4f^15d$ state to a higher ^3P level of the Pr^{3+} ion to the 5d state of the Ce^{3+} ion. This is then followed by the radiative relaxation of the excited 5d electrons of Ce^{3+} ion.

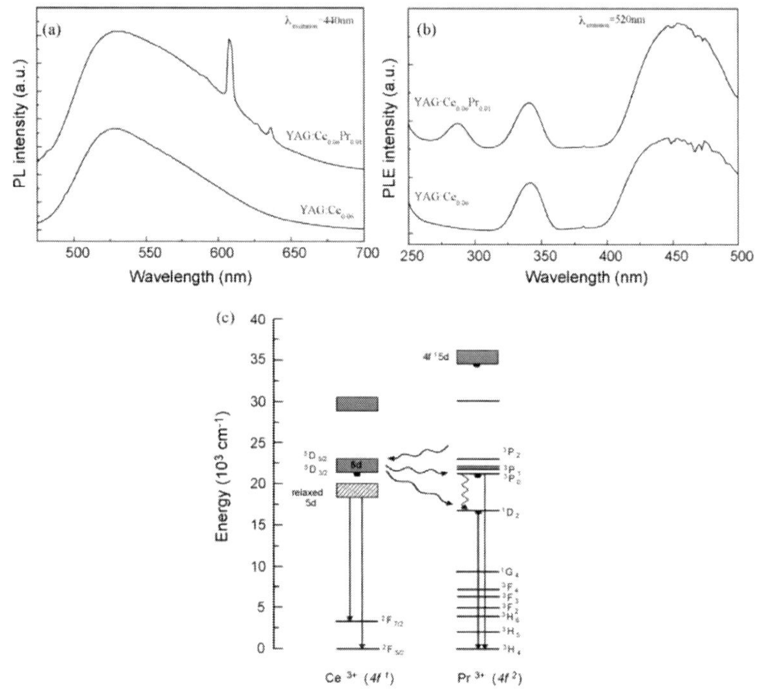

Figure 6.7 (a) PL emission spectra and (b) PL excitation spectra of YAG:Ce and Pr co-doped YAG:Ce samples. (c) Energy transfer processes of Ce^{3+} and Pr^{3+} [21].

In terms of color rendering index, the highest R_a (CRI) of 83 could be obtained from Pr co-doped YAG:Ce phosphor when used in white LED. This was because the red spectral component by Pr co-doping contributed to the improved R_a. However, the luminescence efficiency decreased by about 25% due to the energy transfer process. Pr co-doped Ga substituted YAG:Ce showed a similar R_a value to YAG:Ce based white LED and had very low efficiency, hence is not a good candidate.

(c) Temperature quenching of YAG:Ce phosphors

As the market for high power LEDs is rapidly developing [61, 62], the temperature of the phosphors can reach up to 450 K for commercially available products. In Tamura *et al.*'s work [63], it was found that with increasing temperatures, there was a decrease in emission intensity of blue light from the LED and a red shift in emission maximum. This observation was supported by Sakuma *et al.* [64]. Recently, Bachmann *et al.* [65] provided a better understanding on the effect of temperature quenching on Ce^{3+} emission in YAG:Ce in the high temperature region, particularly in temperatures above 400 K. Such high temperatures have great relevance to high powered white LEDs.

In order to study the luminescence properties of the isolated Ce^{3+} ions, greater experimental focus was placed on the lower Ce concentration sample. This was because for higher Ce concentration samples energy transfers and reabsorption could take place, which would affect the temperature dependence of the luminescence spectra and decay curves. Temperature-dependent luminescence lifetime measurements for low Ce concentration (0.033%) showed that the intrinsic temperature quenching Ce^{3+} was high at well above 700 K. The lower quenching temperatures for commercial YAG:Ce phosphors were explained by thermal-activated concentration quenching, as supported by Robbins *et al.*'s work [66–68]. When lower Ce concentration is used, YAG:Ce is a potentially a good candidate for high power LEDs.

6.4.1.2 Silicate-based phosphor

Due to some limitations of YAG:Ce yellow phosphor, other host materials have been investigated and found to be better candidates for yellow phosphors. Rare earth doped strontium silicate oxides showed promising luminescence properties for use with blue or near-UV LEDs. Strontium silicate nitrate based phosphors have also been reported [69].

6.4.1.2.1 *Luminescence properties of silicate-based phosphors*

Effect of dopants on luminescence properties

Jang *et al.* reported yellow $Sr_3SiO_5:Ce^{3+}$, Li^+ phosphors to be used with blue LED or near-UV LEDs to produce white light. The luminescence excitation spectra of this phosphor showed a strong and broad PLE peak at 415 nm; thus it could be excited by blue (460 nm) or n-UV (405 nm) LED. $Sr_3SiO_5:Ce^{3+}$, Li^+ based white LEDs displayed white light ranging from natural white to cool white without any additional phosphors. The CRI indexes of $Sr_3SiO_5:Ce^{3+}$, Li^+ based white LEDs were larger than the R_a of 60 of white fluorescent lamps. Blue-LED pumped white LED showed excellent bright white light and had especially higher R_a at 82. Near-UV-LED pumped white LED also showed white light with stable color coordinates [70].

Similarly, He *et al.* achieved promising results in their studies on Eu^{3+} doped Li_2SrSiO_4 phosphors, which demonstrated a good yellow light source for white LED applications. Further studies were also performed on Ce^{3+}, Eu^{3+} co-doped Li_2SrSiO_4 phosphor to investigate the luminescence properties and Ce^{3+}-Eu^{3+} energy transfer in the phosphor. The PL spectra of Li_2SrSiO_4: Ce^{3+}, Eu^{3+} phosphor showed two emission peaks in the blue (420 nm) and yellow regime (570 nm), respectively, due to Ce^{3+} and Eu^{3+}, respectively. The energy transfer process was confirmed by PL excitation, UV-vis absorption and PL decay measurements, which showed that there is a spectral overlap between the emission of Ce^{3+} and absorption of Eu^{3+}. With Ce^{3+} co-doping, the absorption of the phosphor with n-UV pumped LED is greatly enhanced [71].

6.4.2 Red-Emitting Phosphors

Red-emitting phosphors are usually combined with green- and/ or blue-emitting phosphors for white LED applications, irradiated under UV light or blue light LED. Research on potential red-emitting phosphors has focused on Eu^{2+} doped materials and sulfides [72, 73]. This is due to the f-f electrons in Eu^{2+}, which corresponds to red-emission wavelength. However, Eu^{2+} doped sulfides and oxides are not favorable due to the low absorption in the blue light range of oxides and low chemical stability of sulfides [18, 74]. Therefore, many recent studies have shown silicon-based nitrides as potentially good host lattices for red luminescence phosphors. Furthermore, many new host materials have been discovered as potential red phosphors for white LED applications and some will be discussed in the following sections.

6.4.2.1 Nitride-based phosphors

Nitridoaluminosilicate and its doped derivatives have received much attention due to their red luminescence properties. (Sr, Ca, Eu)$AlSiN_3$ have been reported [74, 75]. Li *et al.* reported [76] another popular nitride-based phosphor Eu^{2+}-doped $M_2Si_5N_8$ (M = Ca, Sr, Ba), where aluminum was not used. Similar to the nitridoaluminosilicate-based phosphors, $M_2Si_5N_8$ phosphors also showed broadband emission in orange to red spectral range (600–680 nm), depending on the type of metal (M) used and the Eu^{2+} dopant concentration.

6.4.2.1.1 *Luminescence characteristics of nitride-based red phosphors*

(a) Particle size effect on luminescence properties

From the study of Li [77], X-ray diffraction measurements showed that Eu-doped $M_2Si_5N_8$ compounds formed nearly single phase materials. On the fraction x in $M_{2-x}Eu_xSi_5N_8$, it is observed that when M = Ca, the unit cell volume expanded with increasing Eu^{2+} concentration in the range of x = 0–0.14. This was due to the fact that Eu^{2+} (1.17 Å) was larger in size than Ca^{2+} (1.00 Å). When x was greater than 0.14, the unit cell volume remained constant due to the presence of impurity phases. On the contrary, when M = Sr or Ba, the unit cell volume decreased with increasing Eu^{2+} concentration. This was due to the smaller size of Eu^{2+} as compared to Sr^{2+} or Ba^{2+}. $M_2Si_5N_8$ (M = Sr, Ba) experienced crystal lattice shrinkage when Eu^{2+} was added with Ba^{2+} shrinking to a larger extent since Ba^{2+} is larger than Sr^{2+}.

The excitation spectra of $M_2Si_5N_8:Eu^{2+}$ (M = Ca, Sr, Ba) showed excitation peaks at 250, 300, 340, 395, and 460 nm. These excitation peaks were independent of crystal structure, type of M or Eu concentration. The first excitation band at 250 nm was due to the host lattice excitation, while the other excitation bands were due to the $4f^7 \rightarrow 4f^6 5d^1$ transition of Eu^{2+}. The strongest excitation band was around 395 nm; however, with increasing Eu^{2+} concentration, the 5d excitation bands at around 460 nm became more intense with corresponding decrease in dominant excitation band at 395 nm. The efficient excitation in the spectral region of 400–470 nm implied strong absorption as well in this region. This matched closely to the radiative light at around 465 nm of blue LEDs, for example InGaN-based LEDs. This is an important point to note in potential phosphors for white LEDs.

In Mikami *et al.*'s [74] work on nitridoaluminosilicate-based phosphors, the emission peak of Sr rich (Sr, Ca, Eu) $AlSiN_3$ shifted to a shorter wavelength to the orange-red region due to the weaker crystal splitting field resulting from the larger crystal lattice of (Sr, Ca, Eu) $AlSiN_3$ as compared to (Ca, Eu) $AlSiN_3$ since Ca^{2+} ion is larger in size than Sr^{2+} ion. This showed that a smaller particle size resulted in the emission spectrum at a lower energy level. This was consistent with the results of the work from Li *et al.* [77] where the emission peaks of $M_2Si_5N_8$:Eu^{2+} (1 mol%) were at about 605 nm, 610 nm, and 574 nm for M = Ca, Sr, and Ba, respectively. This happens as the particle size of the $M_2Si_5N_8$:Eu^{2+} compound decreases with Ca>Sr>Ba.

(b) Effect of dopant concentration on luminescence properties

Referring to the work of Li *et al.* [76] again, the broad red emission of $M_2Si_5N_8$:Eu^{2+} (M = Ca, Sr, Ba) was due to the $4f^65d^1 \rightarrow 4f^7$ transition of the Eu^{2+} ion. This emission band was strongly dependent on the type of M and Eu^{2+} concentration. For $Ca_2Si_5N_8$:Eu^{2+}, the emission peak shifted from 605 nm to 615 nm as Eu^{2+} concentration increases. The red shift is small due to the limited replacement of Ca^{2+} ion by Eu^{2+}. For $Ba_2Si_5N_8$:Eu^{2+}, there was a larger red shift as compared to $Sr_2Si_5N_8$:Eu^{2+}. $Ba_2Si_5N_8$:Eu^{2+} shifted from orange to the red spectral region, while $Sr_2Si_5N_8$:Eu^{2+} shifted from yellow to the red spectral region as Eu^{2+} concentration increased.

Teng *et al.* [73] proposed $Sr_2Si_5N_8$ as a promising red phosphor and investigated the effects of Eu dopant concentration on luminescence properties of the phosphor. With the increase in Eu^{2+} concentration, the emission band of the phosphor red-shifted from 615 nm to 638 nm under 465 nm excitation. This was due to the increased intensity of the crystal field as more Eu^{2+} ions replaced the Sr^{2+} ions.

(c) Relative luminescence intensities and thermal quenching effect

Li *et al.* further reported that when $M_2Si_5N_8$:Eu^{2+} (M = Ca, Sr, Ba) compounds were excited at 465 nm, $Sr_2Si_5N_8$:Eu^{2+} measured the highest conversion efficiency at 75–80%. The relative luminescence intensity was also the highest at 87%, as compared to 71% (M = Ca) and 84 % (M = Ba) [76]. $Sr_2Si_5N_8$:Eu^{2+} also had the lowest thermal quenching. At 150°C, the intensity of $Ca_2Si_5N_8$:Eu^{2+} has already dropped to 40% of that at room temperature, while $Sr_2Si_5N_8$:Eu^{2+} only decreased by a few percent. Compared with another commercially used red phosphor CaS:Eu^{2+}, at 120°C, the intensity of CaS:Eu^{2+} phosphor was only 80% of that at room temperature, while $Sr_2Si_5N_8$:Eu^{2+} was still of 87% intensity. This shows that $Sr_2Si_5N_8$:Eu^{2+} experiences a comparatively weaker thermal quenching effect and, therefore, has the highest potential as a red-emitting conversion phosphor for white LED applications.

6.4.2.2 Niobate and tantalate

Rare earth niobate and tantalate materials are of great interest in various energy and environmental related applications such as in solid-state lightning, photocatalysts for contaminant degradation and H_2 production, as well as electrolytes in solid-oxide fuel cells [78]. As these materials have various luminescence chromaticities and are chemically and electrochemically stable, rare earth activated niobate and tantalate phosphors have been proposed to be promising materials for optoelectronic applications [79, 80]. However, the poor aqueous solubility and high crystallization temperatures render only solid-state method to be the most suitable synthesis method for niobate and tantalate materials, hence hampering the discovery of new phases, removing impurities and forming of nanoparticles. Therefore, effects of particle size on the luminescence characteristics of niobate and tantalate will not be discussed in the next section.

Further, the different solubilities of rare earth and tantalate oxides also hinder the development of other synthesis methods of rare earth tantalates and niobates [79]. However, hydrothermal synthesis of tantalates [79] and an aqueous synthesis route [78] had been studied and samples showed positive results.

6.4.2.2.1 *Luminescence characteristics of niobate and tantalate red phosphors*

In a recent study [81], the luminescence characteristics of $RNbO_4$:Eu^{3+} (R = La, Y, Gd) incorporated with high concentrations of Bi^{3+} were analyzed and these materials were found to be potential red phosphors for white LED applications. Undoped niobates are known to efficiently emit blue light under X-ray and UV excitation [82–84]. However, when doped with rare earth ions such as Eu^{3+} and Bi^{3+}, the emission peak was red-shifted under UV excitation. The high UV absorption of Eu^{3+} and Bi^{3+} co-doped $RNbO_4$ successfully converts various excitation UV sources into red light.

Similarly, tantalates with no rare earth activator show either weak luminescence at 330–350 nm [120, 121] or no luminescence under UV excitation [78]. Tantalate phosphors are typically doped with Eu^{3+} and Tb^{3+} [85]. Effects of Eu^{3+} doped tantalate phosphor will be described in the (b) below.

(a) Effect of dopants on the luminescence characteristics in niobates

In a study by Park *et al.* on Eu^{3+} and Bi^{3+} co-doped $RNbO_4$ [81], the optimal concentration of Bi^{3+} in Bi^{3+} singly doped $RNbO_4$ was determined first. For Bi^{3+} doped $RNbO_4$, the electron in the 6s orbital of the Bi^{3+} ion could be transferred to the low lying empty 4d orbital of the niobium ion. This resulted in the metal-to-metal charge transfer emitting state. Bi^{3+} doped $RNbO_4$ had two excitation bands, one at 256 nm due to 2p-4d charge transfer transition, and the other at 305 nm due to 6s-4d charge transfer transition. The excitation band at 305 nm was much stronger than the 256 nm band. For $YNbO_4$: Bi^{3+}, the luminescence band peak red-shifted to 448 nm from 410 nm, while the emission peak for $LaNbO_4$: Bi^{3+} remained similar at 410 nm. Bi^{3+} doped $GdNbO_4$ showed the most intense emission peak but did not exhibit luminescence at room temperature. Hence, for $YNbO_4$:Bi^{3+}, luminescence intensity peaked when Bi^{3+} concentration was at 1 atom % while intensity decreased slightly as concentration increased to more than 2 atom %.

As Bi^{3+} was fixed at 0.1 atom % and concentration of Eu^{3+} was gradually increased up to 10 atom %, blue emission from the NbO_4^{3-} group decreased. The blue emission was consumed and resulted in enhancement of the red luminescence at 613 nm corresponding to the Eu^{3+} ion. Following this analysis, the Eu^{3+} activator ion concentration was fixed at 5%, while the Bi^{3+} concentration was gradually increased. The red luminescence of Eu^{3+} was enhanced as Bi^{3+} concentration increased up to 352 nm at half maximum with 15 atom % Bi^{3+} concentrations. A higher Bi^{3+} concentration resulted in decrease in luminescence intensity. Hence, $(Y_{0.80}\ Bi_{0.15})\ NbO_4$:$Eu^{3+}$(5%) may be used as a red-emitting phosphor for white LEDs.

(b) Effect of dopants on luminescence characteristics in tantalates

Nyman *et al.* reported $KLnTa_2O_7$:Eu^{3+} (where Ln = Y, Gd, Lu) [79] to be new promising red-emitting phosphors for white LEDs under blue LED excitation. The PL excitation spectra showed a broad blue excitation band for Ln = Gd especially. This is important in LED applications as a broadband width increases the absorption cross section of the phosphor. The PL emission spectra of the samples showed an intense emission at 608 nm, which is a nearly ideal red emission for high luminous efficacy white LEDs. The quantum yields (QYs) of the $KLnTa_2O_7$:Eu^{3+} phosphors were also high with Ln = Gd close to 80%, and 65% QY peaks for the Ln = Y, Lu. QY was found and confirmed to be dependent on crystalline structure. Nyman *et al.* [78] also synthesized LaTa phosphors of different phases by heating the dry powder samples at different temperatures. The LaTa-850 phase had a QY of 55%, which is much higher than LaTa-1200 phase with

QY of 24.8% under blue excitation with the same Eu^{3+} concentration. The LaTa phosphors also showed red emission luminescence under blue light excitation.

6.4.3 Green-Emitting Phosphors

Currently, the green phosphors used in white LEDs are mostly sulfide based, such as ZnS: Cu^{+}, Al^{3+} [86–88] or $SrGa_2S_4{:}Eu^{2+}$. However, sulfide-based phosphors have limited applications as they have low chemical stability, resulting in strong chemical dependence of chromaticity and poor luminous efficiency in white LEDs. Furthermore, sulfide-based phosphors are unstable toward moisture and other atmospheric components. Thus, there is a need to develop alternative green phosphors for LED applications.

Oxynitrides, nitrides, orthosilicates silicates based green phosphors have been developed for possible white LED applications [89]. Oxyntrides and nitrides based phosphors, mostly activated with Eu^{2+}, have been synthesized. Strong absorption and good PL emission properties under near UV or blue light excitation were reported [18, 74]. However, oxynitrides and nitrides often require harsh synthesis conditions such as high temperature and high pressure [90].

Recently, much attention has been given to orthosilicates or silicates-based green phosphors. Oxide-based host materials like ZnO have also been receiving great research attention in recent years. Hence, the following discussion will focus on silicates and ZnO.

6.4.3.1 Silicates

An important factor in LED phosphors is the ability to tolerate high temperatures of up to 450 K. This is especially important in high power LEDs where the phosphor temperature usually reaches up to 450 K. Alkaline rare earth silicates are suitable as LED phosphor hosts due to their high stability and various crystal structures. Examples include R_2SiO_4 [90–94], $R_2MgSi_2O_7$ [95], $R_2MgSi_2O_8$ [96, 97] where R = Ba, Sr, Sc, Ca, Cl. However, the solid-state reaction method [89, 90, 98] is still commonly used for silicate phosphor synthesis; and, therefore, nanoparticles have rarely been reported for silicate-based green-emitting phosphors. The effect of particle size on luminescence properties will not be discussed here.

6.4.3.1.1 *Luminescence characteristics of silicates*

(a) Effect of dopants on luminescence properties

Divalent europium is an important activator in many luminescent materials. Most green phosphors are activated using Eu^{2+} to give an emission band in the green region. The absorption and emission of Eu^{2+} ions are characterized as many efficient broadband when used in many different hosts as the 4f-5d transition in Eu^{2+} ions are electric-dipole allowed [90].

Zhang *et al.* synthesized $SrBaSiO_4:Eu^{2+}$ [89], which was the intermediate phase of Sr_2SiO_4 and Ba_2SiO_4 by a solid-state reaction method. When excited by 430 nm light, the PLE spectra showed a broad absorption band peaking at 430 nm, attributed to the $4f^7 \rightarrow 4f^6 5d^1$ transition of the Eu^{2+} ion. No red characteristic emission lines due to Eu^{3+} were observed, indicating that all the Eu^{3+} in the crystalline matrix has been reduced into Eu^{2+} ions. The PL spectra showed a wide green emission band, which centers at 512 nm. Under 395 nm excitation, the emission intensity was lowered, but emission profile remained largely similar. The $SrBaSiO_4:Eu^{2+}$ phosphor was tested in a white LED with a 395 nm emitting InGan chip. Figure 6.8 shows the PLE and PL spectra of the observed results. Under a high forward bias current, it was found that $SrBaSiO_4:Eu^{2+}$ has high current tolerability, which renders it useful for LED applications.

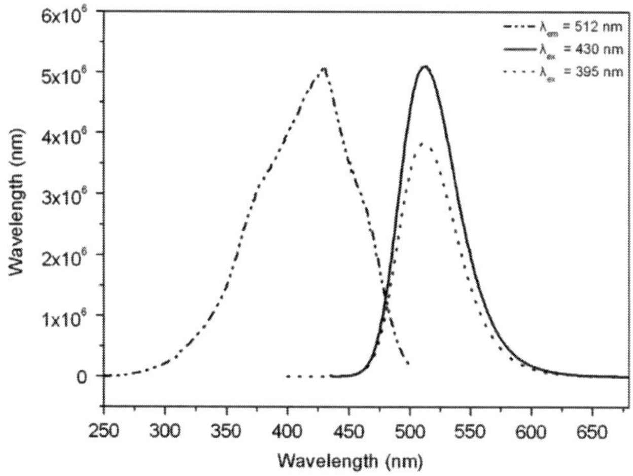

Figure 6.8 PLE (λ_{em} = 512 nm) and PL(λ_{ex} = 395 nm, 430 nm) spectra of $SrBaSiO_4$:0.02Eu^{2+} [89].

Zhang *et al.* [98] also reported similar results when $Sr_2LiSiO_4F:Eu^{2+}$ was synthesized. A broad emission peak was found at 520 nm with an absorption band in the range of 275–450 nm. Short fluorescent decay lifetime measurements showed that it is suitable for solid-state lighting applications. The absorption band matches the near UV wavelength, which allows it to be used as green phosphor in white LED.

Kijima *et al.* reported a cerium-doped green phosphor, $Ca_3Sc_2Si_3O_{12}:Ce$ [99], which gives emission peaks at the blue green region (505 nm) and a green light at 545 nm when excited with blue light. The green light emission corresponded to the transition between the two energy levels $^2F_{7/2}$ and $^2F_{5/2}$ of Ce^{3+}. When used together with another red phosphor and excited by blue LED, this system has been suggested to be promising for white LED applications. When Mg^{2+} replaces part of Sc^{3+}, the emission intensity was found to have increased by 25%.

(b) Effect of dopant concentration on luminescence properties

With an increase in Eu^{2+} concentration, the emission maximum wavelength was shown to be red-shifted and emission intensity was enhanced. Zhang *et al.* [89] reported an optimal emission intensity at Eu^{2+} concentration $x = 0.02$ (molar fraction), after which concentration quenching occurred.

However, in Zhang's study [98], Eu^{2+} was found to have insignificant effect on emission intensity. A possible reason is the energy transfer between Eu^{2+} might not be very efficient. Concentration quenching is dependent on the energy transfer between the nearest neighboring ions in a phosphor system. This has been supported by theoretical calculations by Du *et al.* [90]. Yao *et al.*'s work on a silicate phosphor also proved that the concentration quenching was due to dipole–dipole interaction [100].

(c) Thermal quenching on luminescence properties

Zhang *et al.* [89] reported the quenching temperature of $SrBaSiO_4:Eu^{2+}$ to be 170°C, which was lower than the conventional $YAG:Ce^{3+}$ at 250°C, hence greater decrease in luminescence properties as temperature increased. However, it was significantly better than the currently used green LED phosphors such as $SrGa_2S_4:Eu^{2+}$ and ZnS: Cu^+, Al^{3+}, which have thermal quenching temperatures of 120°C and 160°C, respectively.

The cerium-doped green phosphors such as $Ca_3(Sc,Mg)_2Si_3O_{12}$:Ce and $CaSc_2O_4$:Ce reported by Kijima *et al.* [99] showed weak thermal quenching properties. These phosphors are more suitable for use in white LEDs than the conventional YAG:Ce yellow phosphor, which has a lower thermal quenching temperature.

6.4.3.2 ZnO

Zinc oxide nanostructures have been a subject of great research interest in recent years as they are promising in many optoelectronic applications, such as in LEDs, solar cells [101–103] and buffer layers [2]. They are highly versatile material due to their high transmittance and good electrical conductivity properties [104, 105]. Zinc oxide can also be produced in a variety of nanostructured morphologies under low temperature and low costs [106]. In LEDs, greater emphasis has been placed on ZnO, doped or undoped, to be used as a current spreading layer, electrode, or grown onto GaN-based LEDs to increase light extraction efficiency [107].

In recent years, rare earth doped ZnO nanostructures have been applied in white LEDs and found to be efficient green phosphors for white LED. The effect of dopant on luminescence properties of rare earth doped ZnO will be discussed in the next section.

6.4.3.2.1 *Luminescence properties of rare earth doped ZnO*

(a) Effect of dopant on luminescence properties

Bhatti *et al.* reported co-doping of ZnO with Nd and Pr [108], where the mixture of the rare earth metals Nd and Pr is known as didymium. Didymium-doped ZnO. The samples were excited using a nitrogen laser source of 337 nm, and green emission was observed at 560 nm and 544 nm. Therefore, it is a promising candidate as a green phosphor for white LED applications. Similar emission peaks were achieved by varying dopant concentrations. Furthermore, with the addition of small amounts of a co-activator such as Li^+ ions, the luminous efficiency of the ZnO phosphor was found to be enhanced due to increased energy transfer between the host and rare earth ions.

Ji *et al.* [109] prepared ZnO nanocrystals doped with Tb^{3+}, Dy^{3+}, and Er^{3+}, which resulted in efficient PL in the green region and, hence, are good green phosphors for white LED applications. Tb^{3+} has been studied extensively for its green luminescence properties for various applications [110, 111] and is the most commonly used rare earth dopant in ZnO phosphors. A Tb^{3+}-doped $ZnO\text{-}SiO_2$ binary system has been developed to be an efficient phosphor for LEDs, giving a broad PL emission peak at 524 nm under 325 nm excitation.

The luminescence emission of Tb^{3+}-doped $ZnO\text{-}SiO_2$ was proposed to be due to energy transfer between the host ZnO and the Tb^{3+} ions, as shown in Fig. 6.9. This is supported in another study by Bang *et al.* [122], where they reported an increase in emission intensity of Eu^{3+} ions as more ZnO was incorporated. This suggests that the emission from Tb^{3+} was sensitized by the absorption of excitation energy by ZnO nanoparticles, rather than through the direct excitation of the Tb^{3+} ions. Tb^{3+} ions cannot be directly excited by 320 nm phonons, which suggest a phonon-mediated energy transfer involved or resonance.

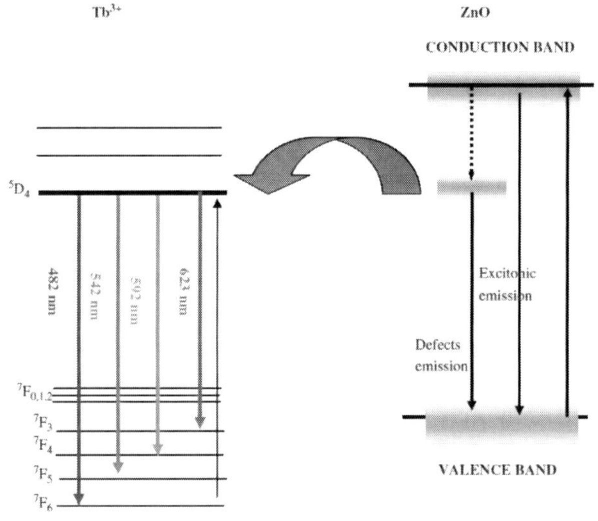

Figure 6.9 Proposed scheme of energy transfer mechanism from ZnO to the Tb^{3+} ions [109].

The presence of ZnO also enhanced the luminescence of the system as it prevented the Tb^{3+} ions from clustering, thus leading to reduced nonradiative de-excitation (which contributes to concentration quenching). This phosphor system gives efficient green emission when excited by light in the near UV range and is a promising candidate for white LED applications.

6.4.4 Blue-Emitting Phosphors

Blue-emitting phosphors are combined with a green- and red-emitting phosphors to produce white light when UV or near UV LEDs are used. A large number of oxide-based phosphors have been reported to emit blue light under UV or near UV excitation. However, these phosphors suffer from high thermal quenching and stability issues. Alternative candidates for blue phosphors, mainly oxynitrides- or nitrides-based phosphors, have been reported recently by Xie *et al.* [18]. Oxynitrides- and nitrides-based phosphors are less prone to thermal quenching and strong absorption in the UV region.

As compared to yellow-, green-, and red-emitting phosphors, there are relatively less reports on the development of blue emitting phosphors. Phosphate based phosphors have recently been found to be a good candidate as blue emitting phosphors for white LEDs.

6.4.4.1 Phosphate-based phosphors

Phosphate-based phosphors have attracted much attention in recent years as they have desirable properties in white LED applications. Liu *et al.* and Wu *et al.* reported the synthesis of $LiSrPO_4$: Eu^{2+} and showed that it has excellent hydrolytic stability and charge stabilization comparable to the commercial blue phosphor: $BaMgAl_{10}O_{17}$:Eu^{2+} (BAM). Other reports showed phosphate-based phosphors to have higher thermal stability, stronger emission intensity, and good phase purity. The luminescence properties of some Eu^{2+} doped phosphate based phosphors will be discussed in the next section.

6.4.4.1.1 *Luminescence properties of phosphate-based phosphors*

The typical PL spectra of phosphate-based phosphors show an excitation band or a few bands in the region between 250 and 440 nm. The emission peak is usually centered at 420 nm [112].

(a) Effects of substituent on luminescence properties

Wan *et al.* [113] compared the emission spectra and intensity of three blue-emitting phosphors with different host matrix due to different substituent on the host: $LiCaPO_4$:Eu^{2+}, $LiSrPO_4$:Eu^{2+}, and $NaCaPO_4$:Eu^{2+}. The PL emission spectrum of $LiCaPO_4$:Eu^{2+} exhibited a symmetrical band between 450 nm and 500 nm with a peak at 470 nm. This peak was due to the $4f^65d^1 \rightarrow 4f^7$ transition of Eu^{2+} that replaced the Ca^{2+} sites in the host lattice. The excitation spectrum showed a single broadband in the 300–450 nm region, as can be seen from Fig. 6.10, where three broad peaks on the left of the graph corresponded to the excitation spectrum of the three samples.

Comparison with the other two phosphors showed that $LiCaPO_4:Eu^{2+}$ had the highest emission intensity, which was about 2.5 times that of $LiSrPO_4:Eu^{2+}$ and 5 times that of $NaCaPO_4:Eu^{2+}$. $LiSrPO_4:Eu^{2+}$ showed blue luminescence, but the emission peak was slightly blue-shifted to 450 nm. Comparing with $LiSrPO_4:Eu^{2+}$ prepared by Wu *et al.* [114], $LiCaPO_4:Eu^{2+}$ was found to have lower sintering temperature for most efficient luminescence and had a higher color rendering index. A blue phosphor with emission peak at 470 nm had the optimum color rendering index [115].

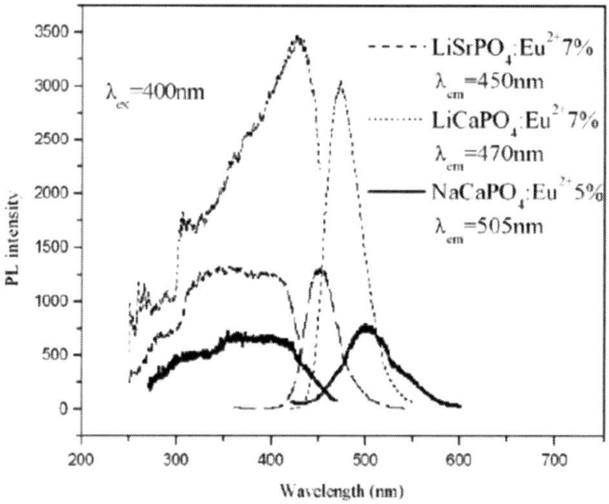

Figure 6.10 Excitation and emission spectra of $LiSrPO_4:Eu^{2+}$, $LiCaPO_4:Eu^{2+}$, and $NaCaPO_4:Eu^{2+}$ [113].

(b) Effects of dopant concentration on luminescence properties

As Eu^{2+} concentration increased, the PL intensity of the blue phosphors reached a peak, followed by a decrease in intensity as critical quenching concentration was reached (Fig. 6.11). The quenching concentration of $LiCaPO_4:Eu^{2+}$ was found to be x (concentration of Eu^{2+}) = 7 mol %. This was almost similar to $LiSrPO_4:Eu^{2+}$ and another phosphate-based blue phosphor reported, $Ca_{1.95}P_2O_7{:}0.05Eu^{2+}$, which was at 5 mol %. Concentration quenching may be understood by looking at the spectra of $LiSrPO_4{:}Eu^{2+}$ phosphor [113, 114, 116]. No emission peaks of Eu^{3+} ions were observed when the concentration of Eu^{2+} was less than 7 mol %; this showed that most Eu^{3+} had been reduced to Eu^{2+}. At Eu^{2+} concentration greater than 7 mol%, it was observed that weak Eu^{3+} emission peak at around 592 nm and 617 nm appeared. This indicated that not all Eu^{3+} were reduced to Eu^{2+} at high dopant concentration. From Fig. 6.13, it can be seen that the emission peak shifts to longer wavelengths with increasing dopant concentration. This might be due to the change in crystal field around Eu^{2+} resulting from slight distortion in crystal structure. The concentration quenching might also be due to energy transfer through Eu–Eu interaction [115].

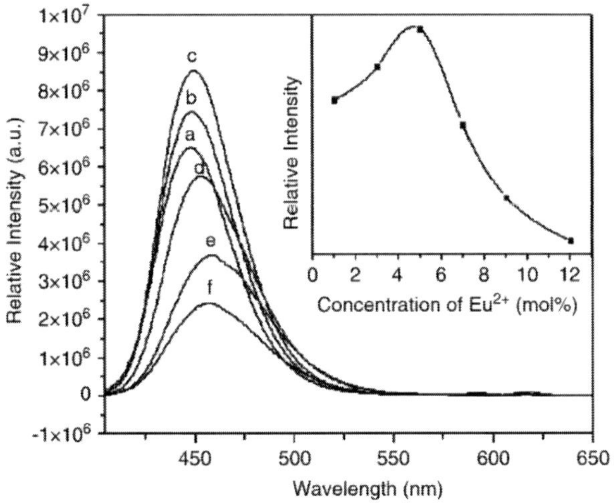

Figure 6.11 $LiSrPO_4{:}Eu^{2+}$ concentration quenching [114].

6.5 Conclusions

Rare earth doped nanophosphors have been widely investigated in recent years for white LED applications, with the potential of replacing conventional bulk phosphors. Rare earth dopants in nanophosphors have shown significant enhancement in luminescence characteristics. Continued effort in rare earth nanophosphors research will deliver higher luminescence performance and cheaper production, which are imperative for commercial LED applications. In particular, more R&D effort is needed for efficient host materials and their simpler large-scale production. The discovery of new rare earth mineral mines and building of new refineries will also help ease rare earth supply crunch and lower the costs of the minerals.

References

1. Zelmon, D. E., *et al.*, (2010) Optical properties of Nd-doped rare earth vanadates, *Applied Optics*, **49**(26), 4973–4978.
2. Bittkau, K., *et al.* (2008) Nanoscale investigation of light-trapping in a-Si:H solar cell structures with randomly textured interfaces, *Physica Status Solidi (A) Applications and Materials*, **205**(12): 2766–2776.
3. Lousteau, J., *et al.* (2010) Tellurite glasses rare earth doped optical fibre devices: recent progress and prospects, In *12th International Conference on Transparent Optical Networks (ICTON), 2010*, doi: 10.1109/ICTON.2010.5549158.
4. Liang, X.; Huang, X.; and Zhang, Q. (2009) Nanophosphors for an enhancement of silicon solar-cell near-infrared response, *Journal of Fluorescence*, **19**(2), 285–289.
5. Verma, R. K., *et al.* (2010) Up and down conversion fluorescence studies on combustion synthesized Yb^{3+}/Yb^{2+}: $MO\text{-}Al_2O_3$ (M = Ca, Sr, and Ba) phosphors, *Journal of Luminescence*, **130**(7), 1248–1253.
6. Wang, F., and Liu, X. (2009) Recent advances in the chemistry of lanthanide-doped upconversion nanocrystals, *Chemical Society Reviews*, **38**(4), 976–989.
7. Zhou, J., *et al.* (2010) Dual-modality in vivo imaging using rare earth nanocrystals with near-infrared to near-infrared (NIR-to-NIR) upconversion luminescence and magnetic resonance properties, *Biomaterials*, **31**(12), 3287–3295.
8. Zhou, J., *et al.* (2010) A versatile fabrication of upconversion nanophosphors with functional-surface tunable ligands, *Journal of Materials Chemistry*, **20**(37), 8078–8085.
9. Kumar, R., *et al.* (2009) Combined optical and MR bioimaging using rare earth ion doped $NaYF_4$ nanocrystals, *Advanced Functional Materials*, **19**(6), 853–859.
10. Xie, R. J., *et al.* (2007) 2-phosphor-converted white light-emitting diodes using oxynitride/nitride phosphors, *Applied Physics Letters*, **90**(19), 191101.

11. Liu, L., *et al.* (2009) Temperature dependent luminescence of yellow-emitting α-sialon: Eu^{2+} oxynitride phosphors for white light-emitting diodes, *Journal of the American Ceramic Society*, **92**(11), 2668–2673.

12. Zhang, Q. Y., and Liang, X. F. (2008) Cooperative downconversion in $Y_3Al_5O_{12}$: RE^{3+},Yb^{3+} (RE = Tb, Tm, and Pr) nanophosphors, *Journal of the Society for Information Display*, **16**(7), 755–758.

13. van der Ende, B. M.; Aarts, L.; and Meijerink, A. (2009) Near-infrared quantum cutting for photovoltaics, *Advanced Materials*, **21**(30), 3073–3077.

14. Smet, P. F., *et al.* (2008) Using rare earth doped thiosilicate phosphors in white light emitting LEDs: towards low color temperature and high color rendering, *Materials Science and Engineering: B*, **146**(1–3), 264–268.

15. Yamamoto, H. (2010) White LED phosphors: the next step, *Proceedings of the SPIE*, **7598**, 759808–759810.

16. Lakshmanan, A., *et al.* (2010) A red phosphor for nUV LED based on (Y, Gd)BO_3:Eu^{3+}, *Materials Letters*, **64**(16), 1809–1812.

17. Kunimoto, T., *et al.* (2009) Recent development of oxyfluoride phosphors for plasma display panels.

18. Xie, R.-J., and Hirosaki, N. (2007) Silicon-based oxynitride and nitride phosphors for white LEDs: a review, *Science and Technology of Advanced Materials*, **8**(7–8), 588–600.

19. Ray, P. C. (2010) Size and shape dependent second order nonlinear optical properties of nanomaterials and their application in biological and chemical sensing, *Chemical Reviews*, **110**(9), 5332–5365.

20. Asefa, T.; Duncan, C. T.; and Sharma, K. K. (2009) Recent advances in nanostructured chemosensors and biosensors, *Analyst*, **134**(10), 1980–1990.

21. Holloway, P. H., *et al.* (2006) Nanophosphors: PL, EL, and biological markers, *Proceedings of 3rd International Symposium on Display and Lighting Phosphor Materials, ISDLPMW'06*, Yokohama, Japan.

22. Yadav, R. S., and Pandey, A. C. (2010) Enhanced efficiency in quantum confined YBO_3:Tb^{3+} nanophosphor, *Journal of Alloys and Compounds*, **494**(1–2), L15–L19.

23. Wang, Z., *et al.* (2010) Low dimensional effects on luminescent properties of $CaWO_4$:Tb nanophosphor, *Journal of the Electrochemical Society*, **157**(4), J125–J129.

24. Pimputkar, S., *et al.* (2009) Prospects for LED lighting, *Nature Photonics*, **3**(4), 180–182.

25. Ponce, F. A., and Bour, D. P. (1997) Nitride-based semiconductors for blue and green light-emitting devices, *Nature*, **386**(6623), 351–359.

26. Khan, A.; Balakrishnan, K.; and Katona, T. (2008) Ultraviolet light-emitting diodes based on group three nitrides, *Nature Photonics*, **2**(2), 77–84.

27. Craford, M. G. (2000) Visible light-emitting diodes: past, present, and very bright future, *MRS Bulletin*, **25**(10), 27–31.

28. Kovác, J.; Jakabovic, J.; and Kytka, M. (2008) Advanced light emitting devices for optoelectronic applications, In *Proceedings of SPIE 2008*, 7138, 71382A.

29. Chun-Hua Yan, C. Z., and Ling-Dong Sun Synthesis of Rare Earth Nanomaterials.

30. Garskaite, E., *et al.* (2010) Luminescent properties of rare earth (Er, Yb) doped yttrium aluminium garnet thin films and bulk samples synthesized by an aqueous sol-gel technique, *Journal of the European Ceramic Society*, **30**(7), 1707–1715.

31. Nien, Y. T.; Chen, K. M.; and Chen, I. G. (2010) Improved photoluminescence of $Y_3Al_5O_{12}$: Ce nanoparticles by silica coating, *Journal of the American Ceramic Society*, **93**(6), 1688–1691.

32. Chen, L.T., *et al.* (2006) Luminescence properties of BAM phosphor synthesized by TEA co-precipitation method, *Journal of Luminescence*, **118**(2), 293–300.

33. Chiang, C. C.; Tsai, M. S.; and Hon, M. H. (2007) Synthesis and photoluminescent properties of Ce^{3+} doped terbium aluminum garnet phosphors, *Journal of Alloys and Compounds*, **431**(1–2), 298–302.

34. Yu-Ling, Y., *et al.* (2010) Co-precipitation synthesis and photoluminescence properties of $(Ca_{1-x-y}, Ln_y)MoO_4$: XEu^{3+} (Ln = Y, Gd) red phosphors, *Journal of Alloys and Compounds*, **505**(1), 239–242.

35. Rongfeng, G.; Wenqing, Z.; and Shuaimou, L. (2009) Preparation and performance of white LED packaged YAG phosphor.

36. Fadlalla, H. M. H., *et al.* (2008) Synthesis and characterization of photoluminescent cerium-doped yttrium aluminum garnet, *Materials Research Bulletin*, **43**(12), 3457–3462.

37. Fadlalla, H. M. H., *et al.* (2008) Synthesis and characterization of single crystalline YAG: Eu nano-sized powder by sol-gel method, *Materials Chemistry and Physics*, **109**(2–3), 436–439.

38. Jia, D., *et al.* (2007) Synthesis and characterization of YAG: Ce^{3+} LED nanophosphors, *Journal of the Electrochemical Society*, **154**(1), J1–J4.

39. Veith, M., *et al.* (1999) Low temperature synthesis of nanocrystalline $Y_3Al_5O_{12}$ (YAG) and Ce-doped $Y_3Al_5O_{12}$ via different sol-gel methods, *Journal of Materials Chemistry*, **9**(12), 3069–3079.

40. Xiao, Q.; Zhou, Q.; and Li, M. (2010) Synthesis and photoluminescence properties of Sm^{3+}-doped $CaWO_4$ nanoparticles, *Journal of Luminescence*, **130**(6), 1092–1094.

41. Huang, J. S., *et al.* (2007) A study of luminescent and structure properties of α-SiAlON:Yb^{2+} phosphors prepared by hydrothermal synthesis process.

42. Ding, Y. S., *et al.* (2007) Structure and luminescent properties of $SrSi_2N_2O_2$ co-doped Yb^{2+} and Nd^{3+} phosphor by hydrothermal synthesis process, Journal/proceedings, volume, page range.

43. Kamiyama, Y., *et al.* (2010) Photostability of YAG: Ce^{3+} nanophosphors synthesized by glycothermal method, *Journal of the Electrochemical Society*, **157**(5), J149–J154.

44. Huang, S. C., *et al.* (2009) Particle size effect on the packaging performance of YAG:Ce phosphors in white LEDs, *International Journal of Applied Ceramic Technology*, **6**(4), 465–469.

45. Xu, S., *et al.* (2009) Effect of fluxes on structure and luminescence properties of $Y_3Al_5O_{12}$:Ce^{3+} phosphors, *Journal of Rare Earths*, **27**(2), 327–329.

46. Shao, Q., *et al.* (2010) Temperature-dependent photoluminescence studies on $Y_{2.93-x}Ln_xAl_5O_{12}$:$Ce_{0.07}$ (Ln = Gd, La) phosphors for white LEDs application, *Journal of Alloys and Compounds*, **498**(2), 199–202.

47. Yang, H. K., and Jeong, J. H. (2010) Synthesis, crystal growth, and photoluminescence properties of YAG:Eu^{3+} phosphors by high-energy ball milling and solid-state reaction, *Journal of Physical Chemistry C*,**114**(1), 226–230.

48. Godlewski, M., *et al.* (2009) Doped nanoparticles for optoelectronics applications, *Low Temperature Physics*, **35**(1), 48–52.

49. Liu, Y., *et al.* (2010) Optical spectroscopy of lanthanides doped in wide band-gap semiconductor nanocrystals, *Journal of Luminescence*, **131**(3), 415–422.

50. Godlewski, M.; Yatsunneko, S.; and Godlewski, M. M. (2007) Light emission properties of nanoparticles doped with transition metal and rare earth ions, In *Proceedings of SPIE 2006*, **6596**, doi:10.1117/12.726353.

51. Godlewski, M., and Leskela, M. (1994) Excitation and recombination processes during electroluminescence of rare earth-activated materials, *Critical Reviews in Solid State and Materials Sciences*, **19**(4), 199–239.

52. Yatsunenko, S., *et al.* (2008) Enhanced energy pumping to Tb^{3+} ions in manganese-doped ZnS nanoparticles, *Journal of Alloys and Compounds*, **451**(1–2), 206–208.

53. Piper, W. W.; DeLuca, J. A.; and Ham, F. S. (1974) Cascade fluorescent decay in Pr^{3+}-doped fluorides: achievement of a quantum yield greater than unity for emission of visible light, *Journal of Luminescence*, **8**(4), 344–348.

54. Fidelus, J. D., *et al.* (2009) Relation between structural properties of Pr^{3+}-doped yttria-stabilized zirconia nanopowders and their luminescence efficiency, *Scripta Materialia*, **61**(4), 415–418.

55. Dicke, R. H. (1954) Coherence in spontaneous radiation processes, *Physical Review*, **93**(1), 99–110.

56. Yang, H.; Lee, D. K.; and Kim, Y. S. (2009) Spectral variations of nano-sized $Y_3Al_5O_{12}$: Ce phosphors via codoping/substitution and their white LED characteristics, *Materials Chemistry and Physics*, **114**(2–3), 665–669.

57. Wu, Z., *et al.* (2009) Preparation of YAG:Ce spheroidal phase-pure particles by solvo-thermal method and their photoluminescence, *Journal of Alloys and Compounds*, **468**(1–2), 571–574.

58. Wei, Z. G., *et al.* (2003) Size dependence of luminescent properties for hexagonal YBO_3:Eu nanocrystals in the vacuum ultraviolet region, *Journal of Applied Physics*, **93**(12), 9783–9788.

59. Kim, D. S., and Lee, R. Y. (2000) Synthesis and photoluminescence properties of (Y, Gd)BO_3: Eu phosphor prepared by ultrasonic spray, *Journal of Materials Science*, **35**(19), 4777–4782.

60. Blasse, G., and Grabmaier, B. C. (1994), *Luminescent Materials*.

61. Robbins, D. J. (1979) Effects of crystal field and temperature on the photoluminescence excitation efficiency of Ce^{3+} in YAG, *Journal of the Electrochemical Society*, **126**(9), 1550–1555.

62. Robbins, D. J., *et al.* (1979) Relationship between concentration and efficiency in rare earth activated phosphors, *Journal of the Electrochemical Society*, **126**(9), 1556–1563.

63. Mueller-Mach, R., *et al.* (2005) Highly efficient all-nitride phosphor-converted white light emitting diode, *Physica Status Solidi (A) Applications and Materials*, **202**(9), 1727–1732.

64. Krames, M. R., *et al.* (2007) Status and future of high-power light-emitting diodes for solid-state lighting, *IEEE/OSA Journal of Display Technology*, **3**(2), 160–175.

65. Bachmann, V.; Ronda, C.; and Meijerink, A. (2009) Temperature quenching of yellow Ce^{3+} luminescence in YAG:Ce, *Chemistry of Materials*, **21**(10), 2077–2084.

66. Robbins, D. J., *et al.* (1979) Temperature dependence of rare earth activated garnet phosphors-1: intensity and lifetime measurements on undoped and *Ce*-doped $Y_3Al_5O_{12}$, *Journal of Electrochemical Society*, **126**(7), 1213–1220.

67. Robbins, D. J., *et al.* (1979) Temperature dependence of rare earth activated garnet phosphors-2: a comparative study of Ce^{3+}, Eu^{3+}, Tb^{3+}, and Gd^{3+} in $Y_3Al_5O_{12}$, *Journal of Electrochemical Society*, **126**(7), 1221–1228.

68. Van Kemenade, J. T. C., *et al.*

69. Seto, T.; Kijima, N.; and Hirosaki, N. (2009) A new yellow phosphor $La_3Si_6N_{11}$:Ce^{3+} for white LEDs.

70. Jang, H. S., and Jeon, D. Y. (2007) White light emission from blue and near ultraviolet light-emitting diodes precoated with a Sr_3SiO_5:Ce^{3+}, Li^+ phosphor, *Optics Letters*, **32**(23), 3444–3446.

71. He, H., *et al.* (2010) $Ce^{3+} \rightarrow Eu^{2+}$ energy transfer mechanism in the Li_2SrSiO_4:Eu^{2+}, Ce^{3+} phosphor, *Optical Materials*, **32**(5), 632–636.

72. Jia, D.; Girardi, J.; and Greenland, K. (2008) Long persistent white light emitting diode.

73. Teng, X., *et al.* (2008) Luminescence properties of nitride red phosphor for LED, *Journal of Rare Earths*, **26**(5), 652–655.

74. Mikami, H. W. *et al.* (2009) New phosphors for white LEDs: material design concepts, *IOP Conference Series: Materials Science and Engineering*, **1**(1), 012002.

75. Uheda, K., *et al.* (2006) Luminescence properties of a red phosphor, $CaAlSiN_3$: Eu^{2+}, for white light-emitting diodes, *Electrochemical and Solid-State Letters*, **9**(4), H22–H25.

76. Hsieh, S.-C., *et al.* (2006) The inhibition of osteogenesis with human bone marrow mesenchymal stem cells by CdSe/ZnS quantum dot labels, *Biomaterials*, **27**(8), 1656–1664.

77. Li, Y. Q., *et al.* (2006) Luminescence properties of red-emitting $M_2Si_5N_8$: Eu^{2+} (M = Ca, Sr, Ba) LED conversion phosphors, *Journal of Alloys and Compounds*, **417**(1–2), 273–279.

78. Nyman, M., *et al.* (2009) Unique $LaTaO_4$ polymorph for multiple energy applications, *Chemistry of Materials*, **21**(19), 4731–4737.
79. Nyman, M., *et al.* (2009) Highly versatile rare earth tantalate pyrochlore nanophosphors, *Journal of the American Chemical Society*, **131**(33), 11652–11653.
80. Arellano, I., *et al.* (2010) Luminescence and structural properties of Y(Ta, Nb)O_4: Eu^{3+},Tb^{3+} phosphors, *Materials Chemistry and Physics*, **119**(1–2), 48–51.
81. Park, T. K.; Ahn, H. C.; and Mho, S. I. (2008) High concentration of Bi^{3+} incorporated into $RNbO_4$: Eu^{3+} (R = La, Y, Gd) as red phosphors for white LED applications, *Journal of the Korean Physical Society*, **52**(2), 431–434.
82. Blasse, G., and Bril, A. (1970) Luminescence phenomena in compounds with fergusonite structure, *Journal of Luminescence*, **3**(2), 109–131.
83. Blasse, G. (1976) Luminescence processes in niobates with fergusonite structure, *Journal of Luminescence*, **14**(3), 231–233.
84. Weitzel, H., and Schröcke, H. Z. (1980) Kristallstrukturverfeinerungen von Euxenit, $Y(Nb_{0.5}Ti_{0.5})_2O_6$, und M-Fergusonit, $YNbO_4$ *Z. Kristallogr.*, 152: p. 69–82.
85. Popovici, E. J., *et al.* (2009) Studies concerning the properties of some europium activated phosphors based on yttrium tantalate.
86. Mills, A. (2005) Phosphors development for LED lighting, *III–Vs Review*, **18**(3), 32–34.
87. Sato, Y.; Takahashi, N.; and Sato, S. (1996) Full-color fluorescent display devices using a near-UV light-emitting diode, *Japanese Journal of Applied Physics, Part 2: Letters*, **35**(7 PART A), L838–L839.
88. Narukawa, Y., *et al.* (2002) Phosphor-conversion white light emitting diode using InGaN near-ultraviolet chip, *Japanese Journal of Applied Physics, Part 2: Letters*, **41**(4A), L371–L373.
89. Zhang, X., *et al.* (2010) An efficient and stable green phosphor $SrBaSiO_4$: Eu^{2+} for light-emitting diodes, *Journal of Luminescence*, **130**(12), 2288–2292.
90. Du, H.; Sun, J.; and Xia, Z. (2009) Luminescence properties of a new green emitting Eu^{2+}-doped barium chlorosilicate phosphor, *Applied Physics B: Lasers and Optics*, **96**(2–3), 459–463.
91. Kim, J. S., *et al.* (2005) Emission color variation of M_2SiO_4:Eu^{2+} (M = Ba, Sr, Ca) phosphors for light-emitting diode, *Solid State Communications*, **133**(3), 187–190.

92. Kim, J. S., *et al.* (2005) Temperature-dependent emission spectra of M_2SiO_4:Eu^{2+} (M = Ca, Sr, Ba) phosphors for green and greenish white LEDs, *Solid State Communications*, **133**(7), 445–448.

93. Hsu, C.-H. Jagannathan, R.; and Lu, C.-H. (2010) Luminescent enhancement with tunable emission in Sr_2SiO_4: Eu^{2+} phosphors for white LEDs, *Materials Science and Engineering: B*, **167**(3), 137–141.

94. Nakano, T., *et al.* (2009) Novel Ba-Sc-Si-oxide and oxynitride phosphors for white LED, *Journal of Luminescence*, **129**(12), 1654–1657.

95. Zhang, M., *et al.* (2007) Luminescence properties of $M_2MgSi_2O_7$: Eu^{2+} (M = Ca, Sr) phosphors and their effects on yellow and blue LEDs for solid-state lighting, *Optical Materials*, **30**(4), 571–578.

96. Im, W. B., *et al.* (2008) Luminescent and structural properties of $(Sr_{1-x}, Ba_x)_3MgSi_2O_8$: Eu^{2+}: effects of Ba content on the Eu^{2+} site preference for thermal stability, *Inorganic Chemistry*, **48**(2), 557–564.

97. Yonesaki, Y., *et al.* (2009) Crystal structure of Eu^{2+}-doped $M_3MgSi_2O_8$ (M: Ba, Sr, Ca) compounds and their emission properties, *Journal of Solid State Chemistry*, **182**(3), 547–554.

98. Zhang, X., and Kim, J. S. (2009) Intense green-emitting Sr_2LiSiO_4F: Eu^{2+} phosphor for n-UV white LEDs, *Applied Physics A: Materials Science and Processing*, **97**(3), 549–552.

99. Kijima, N., *et al.* (2008) New green and red phosphors for white LEDs, *Journal of Light and Visual Environment*, **32**(2), 202–207.

100. Yao, S. S., *et al.* (2009) Concentration quenching of Eu^{2+} in a novel blue-green emitting phosphor: $Ba_2ZnSi_2O_7$: Eu^{2+}, *Applied Physics B: Lasers and Optics*, **96**(1), 39–42.

101. Chen, S., *et al.* (2008) Improved performances in top-emitting organic light-emitting diodes based on a semiconductor zinc oxide buffer layer, *Journal of Luminescence*, **128**(7), 1143–1147.

102. Owen, J., *et al.* (2007) Organic photovoltaic devices with Ga-doped ZnO electrode, *Applied Physics Letters*, **90**(3), 033512-033512-3.

103. Yip, H. L., *et al.* (2008) Polymer solar cells that use self-assembled-monolayer-modified ZnO/metals as cathodes, *Advanced Materials*, **20**(12), 2376–2382.

104. Zhang, T., *et al.* (2006) Influence of ZnO nanorod on the luminescent and electrical properties of fluorescent dye-doped polymer nanocomposite, *Optical Materials*, **29**(2–3), 216–219.

105. Liu, Z., and Li, Y. (2008) Sol-gel synthesis and luminescence property of ZnO: (La, Eu) Cl nanocomposite thin films, *Thin Solid Films*, **516**(16), 5557–5561.

106. Djurišić, A. B., and Leung, Y. H. (2006) Optical properties of ZnO nanostructures, *Small*, **2**(8–9), 944–961.

107. Djurišić, A. B.; Ng, A. M. C.; and Chen, X. Y. (2010) ZnO nanostructures for optoelectronics: material properties and device applications, *Progress in Quantum Electronics*, **34**(4), 191–259.

108. Bhatti, H. S.; Gupta, A.; and Verma, N. K. (2006) Optical properties of doubly doped ZnO phosphors co-activated with lithium ions, *Radiation Effects and Defects in Solids*, **161**(11), 681–685.

109. Dhlamini, M. S., *et al.* (2009) Sensitized luminescence through nanoscopic effects of ZnO encapsulated in SiO_2: Tb^{3+} sol gel derived phosphor, *Physica B: Condensed Matter*, **404**(22), 4406–4410.

110. Yan, B.; Chen, X.; and Wu, J. (2007) Induced assembly and photoluminescence of lanthanum (Tb, Eu, Dy) complexes/ZnO/ polyethylene glycol hybrid phosphors, *Applied Surface Science*, **253**(21), 8575–8580.

111. Monteiro, T., *et al.* (2006) Optical active centres in ZnO samples, *Journal of Non-Crystalline Solids*, **352**(9–20 SPEC. ISS.), 1453–1456.

112. Lin, C. C., *et al.*, (2009) $KBaPO_4$: Ln (Ln = Eu, Tb, Sm) phosphors for UV excitable white light-emitting diodes, *Journal of Luminescence*, **129**(12), 1682–1684.

113. Wan, C., *et al.* (2010) An efficient blue-emitting phosphor $LiCaPO_4$: Eu^{2+} for white LEDs, *Solid State Communications*, **150**(31–32), 1493–1495.

114. Wu, Z.C., *et al.* (2006) A novel blue-emitting phosphor $LiSrPO_4$: Eu^{2+} for white LEDs, *Journal of Solid State Chemistry*, **179**(8), 2356–2360.

115. Fu, D., and Guo, P. M. (2002) China Light and Lighting, **7**, 5.

116. Ta, N., and Chen, D. (2009) Combustion synthesis of β-$Ca_{1.95}P_2O_7$: $0.05Eu^{2+}$ blue phosphor for near ultraviolet excitation, *Journal of Alloys and Compounds*, **484**(1–2), 514–518.

117. Yukio, N. *et al.* (2010) White light emitting diodes with super-high luminous efficacy, *Journal of Physics D: Applied Physics*, **43**(35), 354002.

Index